SK277 Human Biology
Science: Level 2

The Open University

Body Systems

This publication forms part of an Open University course SK277 *Human biology*. The complete list of texts which make up this course can be found at the back. Details of this and other Open University courses can be obtained from the Student Registration and Enquiry Service, The Open University, PO Box 197, Milton Keynes MK7 6BJ, United Kingdom: tel. +44 (0)870 333 4340, email general-enquiries@open.ac.uk

Alternatively, you may visit the Open University website at http://www.open.ac.uk where you can learn more about the wide range of courses and packs offered at all levels by The Open University.

To purchase a selection of Open University course materials visit www.ouw.co.uk, or contact Open University Worldwide, Walton Hall, Milton Keynes MK7 6AA, United Kingdom for a brochure (tel. +44 (0)1908 858793; fax +44 (0)1908 858787; email ouw-customer-services@open.ac.uk).

The Open University
Walton Hall, Milton Keynes
MK7 6AA

First published 2004. Second edition 2006.

Edited and designed by The Open University.

Typeset by The Open University.

Printed in the United Kingdom by Latimer Trend and Company Ltd, Plymouth.

The paper used in this publication is procured from forests independently certified to the level of Forest Stewardship Council (FSC) principles and criteria. Chain of custody certification allows the tracing of this paper back to specific forest-management units (see www.fsc.org).

ISBN 978 0 7492 1878 2

3.2

THE COURSE TEAM

Course Team Chair and Academic Editor

Heather McLannahan

Course Managers

Alastair Ewing

Colin Walker

Course Team Assistants

Catherine Eden

Rebecca Efthimiou

Course Team Authors

Patricia Ash

Pete Clifton

Paul Gabbott

Nicolette Habgood

Tim Halliday

Heather McLannahan

Kerry Murphy

Daniel Nettle

Payam Rezaie

Other Contributors

Vickie Arrowsmith

Leslie Baillie

Production and Presentation Manager

John Owen

Project Manager

Judith Pickering

Editors

Rebecca Graham

Gillian Riley

Bina Sharma

Margaret Swithenby

Design

Sarah Hofton

Jenny Nockles

Illustration

Steve Best

Pam Owen

CD-ROM Production

Phil Butcher

Will Rawes

External Course Assessor

Dinah Gould

Picture Researcher

Lydia Eaton

Indexer

Jane Henley

Course Website

Patrina Law

Louise Olney

SK277 *Human Biology* makes use of material originally produced for SK220 *Human Biology and Health* by the following individuals: Janet Bunker, Melanie Clements, Basiro Davey, Brian Goodwin, Linda Jones, Jeanne Katz, Heather McLannahan, Hilary MacQueen, Jill Saffrey, Moyra Sidell, Michael Stewart, Margaret Swithenby and Frederick Toates.

Cover image: © Alan Schein Photography/CORBIS

CONTENTS

KIDNEY

1.1 Introduction

We have one liver, one heart and one stomach, but two kidneys, one on either side of our bodies (Figure 1.1). We can survive perfectly well with only one functioning kidney; a fact that has led to some poverty-stricken and very desperate people selling, or being coerced to sell, one of their kidneys for organ transplantation (Figure 1.2, overleaf). Indeed, one of the first organs ever offered up for sale on the internet was a kidney, though this practice is now discouraged. The reason why there is a need for 'donated' kidneys is because without at least one functioning kidney, the chances for long-term survival are extremely thin. To give you some indication of the number of people affected and the costs involved in treating kidney disease, here are some figures published by the American National Institutes of Health (comparable figures for the UK are not currently available). In the USA, between 1988 and 1994, over 20 million Americans suffered from some form of kidney dysfunction. In 1998, 3 million people in the USA were told they had kidney disease. Of these, about half a million had kidney failure (end-stage renal disease, ESRD) of whom 72 342 (14%) died. In 2000, the cost of treating ESRD in the USA was estimated to be in the region of 19 billion US dollars.

Healthy kidneys clean the blood by removing excess fluid, minerals and wastes. They also regulate blood pressure (see Chapter 2) and secrete hormones that are involved in the maintenance of healthy bones and

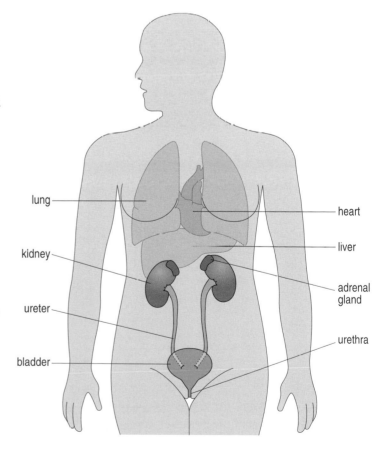

Figure 1.1 Diagram showing the position in the body of the kidneys, ureters, bladder and urethra in relation to other organs.

blood. When the kidneys fail, harmful wastes build up in the body, blood pressure may rise dangerously and the body may retain excess fluid; there may also be a decrease in the production of red blood cells. The aim of this chapter is to describe the key physiological functions of the kidney and to relate them to what happens when the kidneys and their associated structures fail to function normally.

Figure 1.2   Two Moldavian peasants who have each sold one of their kidneys.

1.2 The kidney and homeostasis – fluid balance

In your reading so far you have learnt about the way in which organic macromolecules are utilized to provide energy. You saw that the levels of some molecules in the circulation – most notably glucose – are regulated, and that homeostatic control mechanisms operate to ensure that these levels are maintained within a fairly narrow range. In this chapter, we turn to the regulation of the levels of the simple inorganic substances that are essential to life: water and salts. By now you are familiar with the fact that water is the major component of the intracellular and extracellular fluids, and thereby forms the basis of all body fluids. You are also aware of many of the essential roles played by salts such as sodium and potassium. How the levels of these substances are maintained in the body is the subject of this chapter.

The kidneys play a major role in these homeostatic processes. In particular, the control of water and sodium excretion by the kidneys is crucial in the regulation of body fluid volume and therefore, importantly, of blood pressure, as you will see. However, the kidneys also perform a number of other essential functions, which include the excretion of unwanted or potentially harmful waste products (such as the end-products of metabolism, e.g. urea), the regulation of pH and the production and activation of some hormones. In order to understand these processes, we will need to look at the kidney and its functions at different levels; not simply at the level of the types and arrangement of cells in the kidney, or that of the mechanisms, molecules, and ions involved, but also at the interactions of the kidney with other systems of the body. Indeed, these interactions are fundamental to the roles played by the kidney in homeostasis. We will begin with an overview of the organization and function of the urinary system. In later sections, we will re-examine particular aspects of kidney structure and function in more detail, in order to explain how this organ carries out its essential role in homeostasis.

1.2.1 The basic anatomy of the kidney and the urinary system

As mentioned at the beginning of this chapter, the kidneys are a pair of organs situated at the back of the body, behind the digestive organs (this location is often referred to as being *retroperitoneal*). Urine produced in the kidneys is collected and transported via tubes called **ureters** to the *bladder* where it is stored. Urine leaves the body via the **urethra**. The positions of these organs is shown in Figure 1.1. Collectively, these components are referred to as the urinary system.

If you were to slice a kidney in half, you would be able to see three distinct regions. These are an outer layer, or *renal cortex* (the term 'renal' means 'of the kidney'), a middle layer, or *renal medulla*, and an inner area, or *renal pelvis*, where the ureter widens to join the kidney, as shown in Figure 1.3.

Within each kidney there are around 1 million smaller structures called **nephrons**. Each nephron acts as an independent filter and urine-processing unit (we can think of the nephron as being the functional unit of the kidney). Individual nephrons are too small to be seen in Figure 1.3, but in order to give an idea of the position of nephrons in the kidney, an example has been shown, at an enlarged scale, on the right of the figure. The scale is further enlarged for the diagram of a nephron given in Figure 1.4 (overleaf).

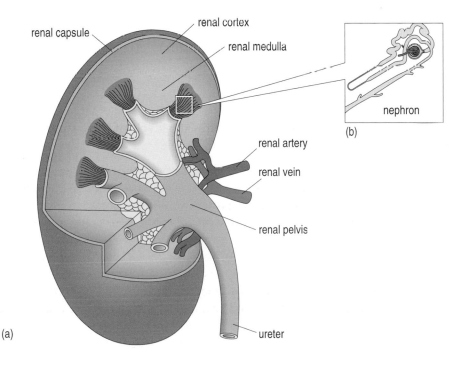

(a)

(b)

nephron

renal capsule

renal cortex

renal medulla

renal artery

renal vein

renal pelvis

ureter

Figure 1.3 Diagrammatic representation of the kidney structure. An enlarged diagram of a nephron is also shown.

Figure 1.4 Larger-scale diagram of a nephron.

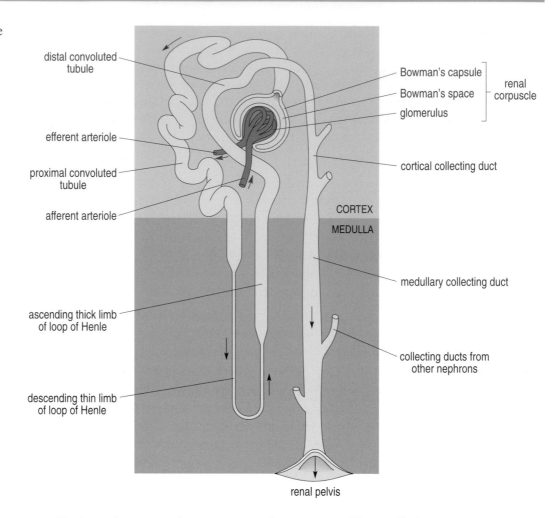

distal convoluted tubule

Bowman's capsule

Bowman's space — renal corpuscle

glomerulus

efferent arteriole

cortical collecting duct

proximal convoluted tubule

afferent arteriole

CORTEX

MEDULLA

medullary collecting duct

ascending thick limb of loop of Henle

collecting ducts from other nephrons

descending thin limb of loop of Henle

renal pelvis

Each nephron contains a structure that acts as a filter, called a **renal corpuscle,** which lies in the cortex, and a long tube which collects and processes the filtered fluid, called a **renal tubule**. At the renal corpuscle, a network of very small-bore blood capillaries, known as the **glomerulus** (plural, *glomeruli*) comes into very close contact with the closed end of the tubule, which is composed of a single layer of epithelial cells. It is at this specialized region of close contact that fluid is filtered out of the blood capillaries, across the epithelial cells and into the lumen of the tubule (recall that the lumen is the internal space within a tubular structure.). The filtered fluid, known as the **filtrate**, then passes along the tubule, which is convoluted (in some cases looping down into the medulla), before finally joining with the renal pelvis, where the urine is emptied into the ureters. It is during its passage along the tubule that the contents of the filtrate are processed, and urine is formed. Most of the filtered water, glucose, amino acids, sodium and other ions are reabsorbed by the epithelial cells of the tubules. Waste substances are either not reabsorbed at all, or only partially reabsorbed. Some molecules and ions are also secreted into the tubule by the epithelial cells and, together with waste products which remain in the filtrate, are excreted in the urine. Although these sound like simple processes, they are closely controlled, so that the levels of water and salts in the body are regulated and unwanted wastes do not accumulate. In order to understand how this control is effected, we will need to look at the processes themselves in more detail.

Summary of Section 1.2

1   The functional unit of the kidney is the nephron; each of the two kidneys contains around one million nephrons.

2   Each nephron consists of a renal corpuscle, where fluid is filtered out of the blood, and a renal tubule, where the composition of the filtrate is modified, mainly by reabsorption of 'wanted' substances.

3   The processed filtrate forms the urine, which passes from the kidneys to the bladder via the ureters.

1.3 Processes involved in the formation of urine

Three main processes are involved in the formation of urine by the kidneys. These are *glomerular filtration*, *tubular reabsorption* and *tubular secretion*. Some metabolic reactions, which have an indirect effect on the formation of urine, also take place in the tubules. We will examine these different processes in turn.

1.3.1 Glomerular filtration

The first stage in the formation of urine is the filtration of fluid from the blood at the renal corpuscle. Blood enters the kidney via a blood vessel called the **renal artery** (see Figure 1.3) and travels through vessels of decreasing size to the renal corpuscle, where the **afferent arterioles** (i.e. those carrying blood *towards* the nephron; Figure 1.4) give rise to the capillaries of the glomerulus. Surrounding the glomerulus is the closed end of the tubule, which forms a hollow cup-like structure, known as **Bowman's capsule**. This arrangement is shown in Figure 1.5.

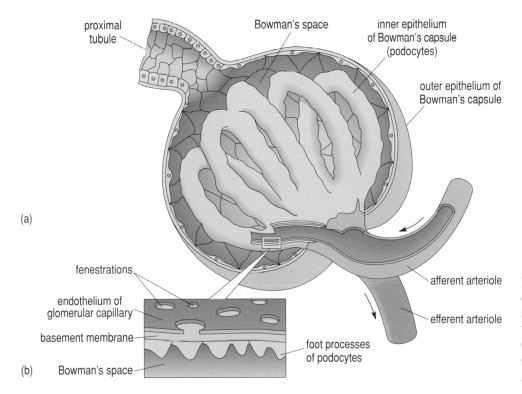

Figure 1.5 (a) Simplified diagram of the renal corpuscle, showing the glomerular blood vessels and Bowman's capsule. (b) Enlarged diagram showing cellular specializations at the glomerulus.

During passage through the glomerulus, the pressure of the blood in the capillaries forces some water, small molecules and ions across the capillary and tubule walls, into the space in the middle of the closed end of the kidney tubule, known as *Bowman's space*. One way to visualize the arrangement of the corpuscle is to imagine a clenched fist (the glomerulus) pushed into the top of an inflated balloon (Bowman's capsule). The outer surface of the top of the balloon would be in close contact with the fist. This is also the case in the renal corpuscle – the epithelial cells of the inside of Bowman's capsule are in close contact with the endothelial cells that form the walls of the capillaries (endothelial cells form a single layer of cells, often referred to as the **endothelium**, that line the walls of the heart, blood vessels and lymphatic vessels). Fluid containing small molecules and ions is forced through the walls of the capillaries and of Bowman's capsule, into the tubule (in our analogy, this would be equivalent to the inside of the balloon).

The fluid filtered from the blood then passes along the tubule, via the renal pelvis, to the ureters; the unfiltered blood remaining in the capillaries leaves the glomerulus via the **efferent arterioles**.

The process of glomerular filtration is crucial to kidney function. There are several special features of the arrangement at the glomerulus that enhance the rate of filtration. The pressure at the glomerular capillaries is much higher than that of other capillary beds because blood in the afferent arteriole supplying the glomerulus is at a higher pressure than that in other arterioles (Section 2.5.5). Also, the arrangement and properties of the cells of the glomerulus are different from those of capillaries elsewhere in the body. This is not only because the walls of the glomerular capillaries and the Bowman's capsule are in close apposition, but also because the cells of the capillaries and capsule have some special structural properties. The endothelial cells lining these capillaries are unusual; they are extremely thin and have many pores called fenestrations (small holes), which are large enough to allow small molecules to pass through, but too small to allow filtration of very large molecules (macromolecules) or, of course, blood cells. The basement membrane (see Figure 1.5b), a gel-like extracellular material secreted by the cells of the glomerulus, confers a degree of electrical selectivity on the filtration process. Most of the proteins in the basement membrane are negatively charged glycoproteins; they repel other negatively charged macromolecules and hinder their passage into the tubule. Because most proteins in the blood (plasma proteins) also bear an overall negative charge, this electrical repulsion acts together with the molecular sieve provided by the fenestrations to slow or prevent their entry into the tubule. The epithelial cells of the inside of Bowman's capsule, which lie next to the capillaries are also unusual; they have foot-like processes (extensions) which leave spaces or channels, through which fluid can pass. The unusual shape of these cells has led to their being named *podocytes* ('foot cells'). The specializations of cells of the glomerulus are shown in Figure 1.6.

The liquid part of the blood is called *plasma* (described fully in the next chapter) and contains many different substances. Around 20% of the plasma that enters the glomerulus is filtered out of the capillaries into Bowman's space. The liquid which passes out of the blood vessels, which is known as the *glomerular filtrate*, has nearly the same composition as plasma, except that it does not contain proteins. Another difference between plasma and the filtrate is due to the fact that some small molecules and ions remain bound to proteins in the plasma, and so these too,

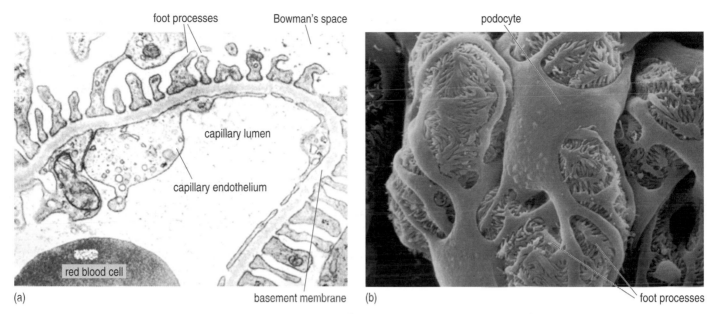

Figure 1.6 (a) Electron micrograph of part of a glomerulus. (b) False-colour scanning electron micrograph of podocytes showing the foot-like processes.

are not filtered. Examples include plasma calcium, of which around half is bound to protein, and fatty acids, virtually all of which are in the form of lipoproteins. Table 1.1 lists the constituents of the glomerular filtrate and the surrounding glomerular capillaries.

Table 1.1 Constituents of glomerular filtrate and glomerular capillaries.

Blood constituents in glomerular filtrate	Blood constituents remaining in the glomerular capillaries
water	water
mineral salts	white cells
amino acids	red blood cells
ketone bodies (fatty acid metabolites)	platelets
glucose	plasma proteins
hormones	some drugs
creatinine (waste product from muscle)	
urea (main nitrogenous waste molecule from protein)	
uric acid (nucleic acid waste product)	
toxins	
some drugs	

The volume of fluid filtered per unit time is defined as the **glomerular filtration rate (GFR)**. For a 70 kg individual, this rate is around 180 litres per day, or 125 millilitres (ml) per minute. This contrasts with the total amount of fluid that filters out into interstitial fluid from all the other capillaries in the body, which is around four litres per day. You may be interested to compare these values with those of the total blood volume, which of course varies between individuals but is about five litres, and the total plasma volume, which is about three litres. The massive rate of filtration by the kidneys is achieved not only because of the cellular specializations at the glomerulus, but also because the kidneys receive a large proportion (20–25%) of the output from the heart.

● You have seen that, during filtration, fluid is forced through the blood capillary and tubule walls at the glomerulus; what factors do you think would influence the rate of this process?

● The rate of filtration is influenced by the pressure of the blood in the afferent and efferent arterioles, and by the permeability of the walls of the glomerulus and Bowman's capsule. (The relative concentrations of proteins in the glomerular capillaries and Bowman's space also have an effect on GFR.)

It may help to think of a hose, leaking because of a number of holes. The greater the number and size of the holes, the greater the rate at which water leaks out. However, factors such as pore size and number, and the composition of the extracellular matrix in the glomerulus do not undergo rapid or significant changes under normal circumstances (although they may be affected in some disease states). Returning to the hose, water will also leak faster if the tap is turned on further, increasing the water pressure. A similar effect would be achieved by squeezing the hose beyond the holes. Similar mechanisms operate in the kidney; GFR is affected by factors that affect blood pressure in the afferent and efferent arterioles. We will return to this later.

There is another factor affecting the filtration rate in the kidney, which is not so easily compared with our analogy of the hose; this is the relative concentration of proteins in the glomerular capillaries and in the Bowman's space. As water and small molecules and ions filter out of the blood, so the protein concentration in the glomerular capillaries increases and osmotic pressure tends to 'pull' the water back into the capillaries (osmosis was described in Book 1, Box 4.4). This exerts a negative effect on filtration rate. However, this influence on filtration rate is less significant than that of blood pressure in the glomerulus, which is high.

Under normal circumstances, GFR does not vary greatly despite changes in mean arterial blood pressure (i.e. pressure in arteries (Book 1, Section 1.1), *not* arteriolar pressure, mentioned above). We will return to why this is so when we look at the control of GFR later in this chapter.

The total amount of any substance filtered through the glomerulus (i.e. *not* proteins or protein-bound substances) per unit time can be measured by determining the plasma concentration of the substance, and multiplying it by the glomerular filtration rate. The resulting value is called the **filtered load** of the substance.

● If the concentration of substance X in the plasma is 0.8 mg (milligrams) per ml, and the filtration rate is 125 ml per min, what will be the filtered load of substance X?

● The filtered load of substance X will be $0.8 \times 125 = 100$ mg per min.

This is the maximum rate at which a substance can be removed from the body. This value is *not* likely to be the same as the rate at which the substance is excreted. This is because the composition of the filtrate is modified as it passes along the tubules; this will be covered in more detail in the following sections. However, before we move on we first need to revisit the anatomy of the nephron.

The first part of the tubule, known as the **proximal convoluted tubule**, is where most reabsorption of glucose, amino acids and water occurs. After passing through the proximal tubule, the filtrate reaches an area known as the **loop of Henle** (pronounced 'henlee'). The epithelial cells forming the first part of the loop are much thinner than those of the proximal tubule (see Figure 1.7) and they have no microvilli and few mitochondria.

The first part of this so-called thin region of the loop of Henle, called the descending limb, is a site of passive movement of water, and to a lesser extent, of ions. The epithelial cells in this region are permeable to these substances and the filtrate is concentrated by the reabsorption of water by osmosis, since the osmolarity of the

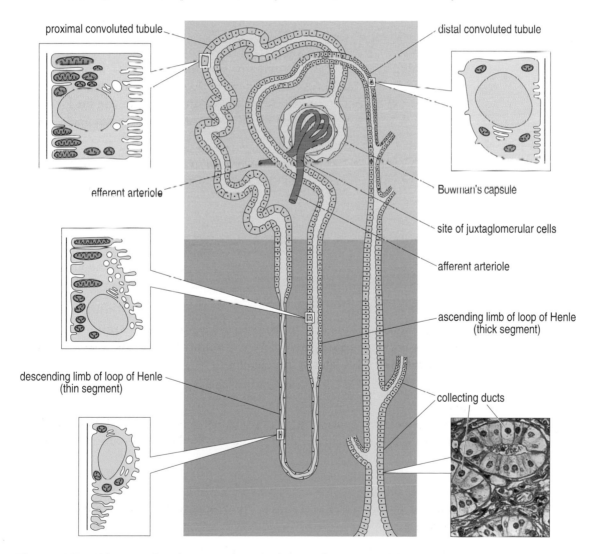

Figure 1.7　Diagram showing a nephron and the main structural features of epithelial cells in the different parts of the tubule.

surrounding tissue is greater than that of the filtrate (you will learn more about this in Section 1.5.1). However, the second thin part of the loop of Henle, the ascending thin limb, is less permeable to water.

The thickness of the tubule wall increases about half-way up the ascending limb; the epithelial cells here are much deeper, and contain more mitochondria (Figure 1.7). Na^+ and Cl^- ions are absorbed by active transport in this region of ascending limb. Other ions, particularly K^+ and H^+, are secreted into the tubule in this area. Upon returning to the region of the renal corpuscle, the ascending limb shows an extremely important anatomical arrangement, which is of fundamental importance in the control of kidney function. If you look at Figure 1.8, you will see that the tubule passes between the afferent and efferent arterioles. In fact, the tubule passes extremely close to the glomerulus. We will return to this point in Section 1.4.1.

After the tubule passes close to the glomerulus, it is known as the **distal convoluted tubule**, which empties into the **collecting duct system**. The final control of urine composition occurs by reabsorption and/or secretion in this part of the nephron. Collecting ducts from several nephrons merge, and the filtrate is emptied into the renal pelvis.

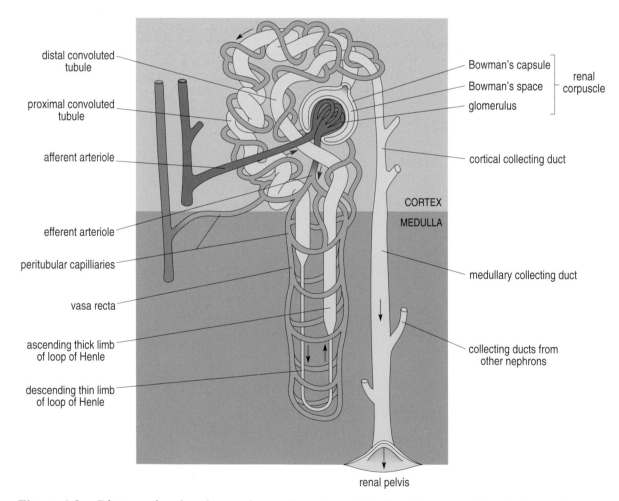

Figure 1.8   Diagram showing the renal vascular system. Note the close association between the peritubular capillaries and the nephron.

At this stage, you may be wondering how the reabsorbed molecules are returned to the rest of the body. The answer is that there is another capillary network in the kidney, in addition to the glomerular system. This is known as the **peritubular capillary system**, and is illustrated in Figure 1.8 which shows how the capillaries surround the tubular system, allowing passage of reabsorbed substances back into the blood (substances that are secreted by the tubular epithelium pass *out* of the blood in this region). A loop of blood vessels, known as the *vasa recta*, follows the path of the long loops of Henle.

1.3.2 Tubular reabsorption

After filtration at the glomerulus, the filtrate passes out of Bowman's space into the main part of the tubule, the walls of which are also composed of a single layer of epithelial cells. You will remember that the tubule has a complicated anatomy, looping down and back up again (Figure 1.7). During the filtrate's passage through the tubule, its composition is modified, predominantly by processes known collectively as *tubular reabsorption*, by which substances pass back into the blood across the epithelium.

It is important to realize that, while the amounts of water and many ions that are reabsorbed are closely controlled in order that their levels in the body are regulated, not all reabsorption from the tubules is controlled. The reabsorption of organic nutrients, such as glucose, many amino acids and water-soluble vitamins, for example, is actually not controlled at all. This is because, under normal circumstances, virtually all of these physiologically useful organic molecules are reabsorbed, and so their levels in the circulation are not changed, but maintained, after passage through the kidney.

To understand *how* the reabsorption of water and salts is controlled, it is necessary to look in more detail at the reabsorption processes involved, and also at the organization of the nephron. First, we will consider how different molecules and ions are reabsorbed from the tubules.

Reabsorption of glucose and amino acids

Glucose and amino acids are absorbed by the epithelial cells of the kidney tubules in the same way that they are absorbed from the gut, i.e. by active transport, against a concentration gradient. In the tubule, as in the gut, glucose and amino acids are cotransported with sodium ions (see Figure 1.9a, overleaf). It is the concentration difference of sodium across the epithelium which is the direct driving force for the active transport of glucose and amino acids. This gradient is set up by the action of the **sodium pump** (also known as the sodium-potassium ATPase), the mechanism of which was first described in the caption of Figure 4.16 in Book 1. As in the absorptive epithelium of the gut, in the kidney tubules the sodium pumps are localized on the basal and lateral sides of the epithelial cells (i.e. on the sides that are *not* facing the lumen). Consequently, sodium ions are pumped out of the cells into the interstitial fluid between them. This results in a lowering of the concentration of sodium inside the epithelial cells, so that it is less than the concentration in the lumen of the tubules. This gradient allows the active transport of sodium, together with a cotransported molecule such as glucose, into the epithelial cells from the lumen of the tubules.

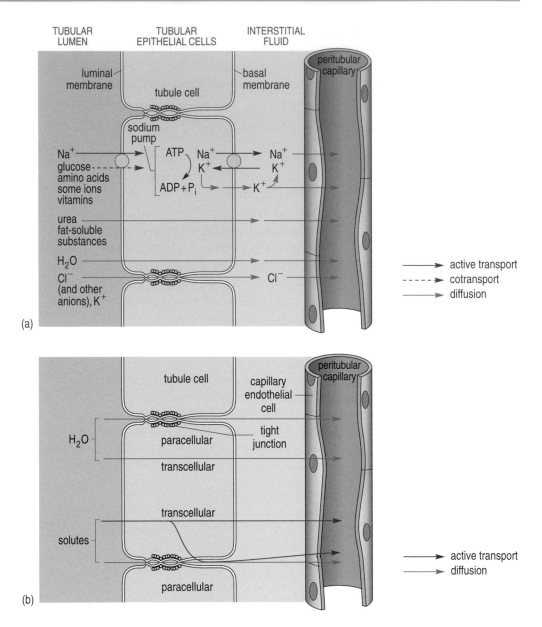

Figure 1.9 Schematic representation of tubular reabsorption. (a) Reabsorption by the proximal convoluted tubule. Most often sodium (Na^+) entry at the luminal surface is coupled to the cotransport of other molecules, such as glucose and amino acids. After sodium enters the cell it diffuses to the basal membrane where it is removed and deposited into the interstitial fluid by the action of the sodium pump. From there it diffuses into the peritubular capillaries. Active pumping of sodium at the basal membrane creates the concentration and osmotic gradients that drive the reabsorption of water by osmosis, of chloride (and other negatively charged molecules) and fat-soluble substances by diffusion, and of organic nutrients and selected positively charged molecules by active transport. (b) Water is reabsorbed by both the paracellular route (through tight junctions; see Book 1, Chapter 4) and the transcellular route. This process is very similar to the reabsorption of water seen in the gut (Book 1, Figure 4.16).

When the levels of glucose and amino acids in the blood are within the normal range, these nutrients are reabsorbed completely during passage through the kidney.

● There is, however, a limiting factor in the reabsorption of glucose or amino acids. Can you work out what this is?

● The number of transporter molecules in the luminal membrane of the tubular epithelial cells.

When blood glucose levels are high, as occurs in diabetes mellitus (Book 2, Section 3.7.2), then more glucose will be filtered, and the concentration of glucose in the filtrate will be increased. Under these circumstances, there can be too much glucose for the transport proteins to handle; the transport proteins may be constantly in use, so that not all the glucose present in the filtrate is able to bind to a free transport site. Another way to describe this is that the transporter molecules are *saturated*, that is, all the sites on the transporter proteins which bind glucose may be filled. Under these circumstances, the excess glucose is not reabsorbed, but is excreted in the urine, making it sweet (hence the name *mellitus*, from the Latin word for honey). Diabetes mellitus is caused by either a reduction in the release of insulin, the hormone that regulates blood glucose levels, or a decrease in the number and/or sensitivity of the insulin receptors (often caused by autoimmune disease as will be described in Chapter 4).

Given the growing incidence of diabetes mellitus in the general population (currently 3–5% and projected to grow to more than 10% as a consequence of increasing levels of obesity and lifestyle choices) it is very likely that at one time or another your doctor has asked you to provide a urine sample for glucose testing. Figure 1.10 shows a urine sample which has been tested using a simple dip-strip to detect the presence of glucose.

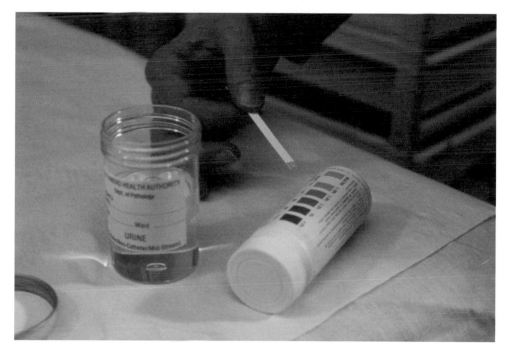

Figure 1.10 Test for glucose in urine. The blue colouration on the end of the reagent strip indicates a negative (normal) result. Glucose levels in urine are raised in diabetes mellitus.

Reabsorption of other organic molecules

Most other useful organic molecules, such as the vitamins, are also reabsorbed by active transport in the first part of the renal tubule. If an excess of water-soluble vitamins, such as vitamin C, is ingested, the excess is also excreted, like an excess of glucose, because of the saturation of transport proteins. However, this is not the case for lipid-soluble vitamins (such as vitamins A and D) and other lipid-soluble chemicals, which can be toxic at high levels. An excess of lipid-soluble molecules in the filtrate tends to diffuse passively into the epithelial cells lining the renal tubule; so, if ingestion continues, the levels of these molecules in the body rise.

● Why is it that lipids can passively diffuse into the epithelial cells and amino acids cannot?

● Lipids pass through the plasma membrane because they are *lipophilic* molecules; it should be remembered that a major constituent of the plasma membrane is lipid. Hydrophilic molecules, such as amino acids, do not readily pass through the lipid layer of the membrane, hence the need for carrier cotransporter proteins to shuttle them across it.

Such potentially toxic substances are, in some cases, modified (in the liver) to make them less-lipid soluble, and thus more readily excreted by the kidneys. They may also be excreted in bile. However, these modification processes are not totally efficient, so excretion remains incomplete, allowing accumulation of these substances in the body. This is one reason for the toxicity of chemicals such as the pesticide DDT and *high* levels of vitamins A and D.

Organic molecules that are not useful, for example the end-products of some metabolic reactions, may also be partially reabsorbed by diffusion, although most are excreted. The main metabolic waste product in plasma is urea, which is a nitrogen-containing molecule formed during the catabolism of amino acids. Urea is a small molecule which is filtered at the glomerulus and is passively absorbed in the tubules (Figure 1.9a). Around 40–60% of the urea that is filtered is reabsorbed, the remainder is excreted in the urine. Another important molecule excreted in the urine is creatinine, a waste product of skeletal muscle metabolism. Unlike urea, creatinine is not reabsorbed in the tubules.

Reabsorption of sodium

Sodium is reabsorbed both by active transport and by facilitated diffusion via protein carriers in the cell membranes of the tubular epithelial cells. Transport of sodium is not only coupled to the cotransport of glucose and amino acids as described above; several other types of sodium transport molecules are also present in the epithelium of the kidney tubules. These transport other ions, such as bicarbonate or phosphate, together with sodium. Sodium reabsorption is also coupled to transport of hydrogen ions *out* of the epithelial cells by some cotransport proteins. Ultimately, all absorption of sodium – by both active transport and diffusion – is dependent upon the decreased concentration of sodium inside the tubular epithelial cells, a result of the action of the sodium pump (see Figure 1.9a).

The mechanisms by which sodium enters the epithelium varies according to the region of the tubule; this is because the epithelial cells of different regions have different transporter proteins. Thus, different substances are cotransported with sodium in different parts of the nephron. For example, glucose is only reabsorbed in the first part of the tubule, while chloride and potassium are cotransported with sodium further along the tubule.

Reabsorption of ions other than sodium

You have seen how many ions, such as phosphate, are absorbed, with sodium, by active transport, and that the absorptive properties of the tubular epithelium vary in different parts of the nephron. Other ions are also absorbed by diffusion and, in the case of chloride ions, which as you will remember are negatively charged, by passive movement with the positively charged sodium ions.

Reabsorption of water

As in the gut, the absorption of water in the kidney tubules is a passive process, which occurs by osmosis (see Book 1, Box 4.4) and diffusion, because of gradients in the concentrations of other substances across the epithelium, most particularly sodium. You have seen that the action of the sodium pump creates a gradient of sodium concentration, which ultimately drives the active transport of many different molecules and ions. The reabsorption of sodium and these other substances has the effect of diluting the filtrate and raising the solute concentration of the surrounding tissue. Thus water tends to pass passively, by osmosis, out of the lumen into the epithelial cells and by diffusion through the junctions *between* the epithelial cells, which, in some parts of the tubule, are very leaky (Figure 1.9b).

Now, we need to pause for a moment to consider the process of osmosis in a little more detail. You have learnt that osmosis is the passage of water across a semipermeable membrane from a region of low solute concentration to a region of high solute concentration, and that this process is important both in the absorption of water from the gut, and the reabsorption of water in the kidney. The osmotic strength of a solution is expressed as its *osmolarity*, which is the number of solute particles in a unit volume of solution. A solution that exerts a high osmotic pressure contains a large number of solute particles, and has a high osmolarity. Because osmotic pressure is dependent upon the *number* of particles in solution, substances that dissociate into ions in solution exert a higher osmotic pressure than those that do not, since they will contain more osmotically active particles (e.g. NaCl will dissociate into two ions: one Na^+ ion and one Cl^- ion).

Water is absorbed from the filtrate by osmosis mainly as a result of sodium reabsorption. However, a crucial point is that water can only pass across the epithelium *if* the epithelium is permeable to water; in fact the properties of the tubular epithelium vary along the length of the tubule, so that, for example, while water is readily reabsorbed in the proximal tubule, it is not absorbed so readily in some other regions, as you will see.

1.3.3 Tubular secretion and metabolism

During passage through the tubules, while some molecules and ions are removed from the filtrate by reabsorption, others are *added* to the filtrate. This occurs by secretion from the tubular epithelial cells, into the lumen of the tubules. The main ions secreted are hydrogen (H^+) and potassium, (K^+) by active transport coupled to sodium reabsorption. Some products of metabolism (metabolites) such as creatinine, and some ingested organic molecules, for example drugs such as penicillin, are also secreted into the tubules. This is one way in which molecules that are not filtered at the glomerulus can be excreted. Both active transport and facilitated diffusion are involved in secretion at the tubules. In addition to absorption and secretion, the tubule cells also carry out some important metabolic processes, which are given the general term *tubular metabolism*.

1.3.4 Renal clearance

From the previous section, you should now realize that the amount of a substance excreted in the urine is a result of the combination of the processes of filtration, reabsorption and secretion. This can be expressed, for a substance X, by the equation:

amount of X excreted =

(amount of X filtered + amount of X secreted) – amount of X reabsorbed

The overall fate of substances in the tubule can be determined by measuring the amount present in the urine, and comparing this value with that of the filtered load. From this we can determine whether a substance undergoes net reabsorption (i.e. is not excreted) or net secretion (i.e. is excreted).

For clinical purposes, renal function is measured by determining the clearance of substances from the plasma. The **renal clearance** of a particular substance is defined as the volume of plasma from which the substance is completely cleared from the kidney per unit time.

● If a substance is neither secreted nor reabsorbed what is its renal clearance?

● The renal clearance of that substance will be equal to the glomerular filtration rate.

In fact, there is no substance occurring naturally in the body that has these properties. However, one compound in the body that is not reabsorbed and undergoes only very little secretion is creatinine; this substance is therefore usually measured in order to assess the effectiveness of glomerular filtration.

So, in order to measure clearance of any substance, the amount of that substance in the plasma, and the amount excreted in the urine per unit time must be measured:

$$\text{clearance of X/ml per min} = \frac{\text{amount of X excreted/mg per min}}{\text{plasma concentration of X/mg per ml}} \qquad (1.1)$$

● What will be the clearance of urea if the plasma concentration is 0.26 mg per ml and the amount excreted is 18.2 mg per min?

● The clearance of urea will be 18.2/0.26 = 70 ml per min.

● What will be the clearance of glucose, which has a plasma concentration of 0.8 mg per ml, and is completely reabsorbed?

● Zero. No glucose is cleared from the kidney under normal circumstances (i.e. in the above equation, 0/0.8 = 0!)

The composition of urine is listed in Table 1.2.

Table 1.2   The composition of urine.

Composition of urine	
water	96%
urea	2%
uric acid	
creatinine	
ammonia	
sodium	
potassium	2%
chlorides	
phosphates	
sulfates	
oxalates	

The processes just described explain how molecules and ions are filtered, reabsorbed and secreted, but offer no explanation of how these processes are controlled. What happens, for example, if excess fluid is drunk as can happen when drinking socially, or a very salty meal is ingested? There are clearly great variations in the intake of these substances, and in their loss from the body by other routes, for example during sweating. There are times when it is crucial that not all sodium and water are reabsorbed, but others when it is essential to conserve every last drop of water. In the following section we will see how renal processes are controlled, so that water and sodium levels, and also the levels of several other important variables, can be regulated.

Summary of Section 1.3

1   The main processes involved in the formation of urine are glomerular filtration, tubular reabsorption and tubular secretion.

2   Plasma is filtered out of the blood at the renal corpuscles, where the glomerular capillaries are surrounded by the epithelial cells of a special part of the kidney tubules, known as Bowman's capsule.

3   The high rate of filtration of plasma at the renal corpuscles is made possible by the high pressure of the blood in the glomerular capillaries, and by structural specializations of both the capillary endothelial cells and the tubular epithelial cells, and by the chemical properties of the extracellular matrix which lies between them.

4   Large molecules, such as proteins, and also protein-bound molecules, such as fatty acids, are not filtered at the renal corpuscle.

5   Both the structure and the transporter proteins of the tubular epithelial cells vary in different regions of the tubule; the substances reabsorbed in different parts of the tubule therefore also vary.

6   After filtration, the filtrate is processed as it passes along the tubule. Physiologically useful molecules, such as glucose, amino acids and many ions are reabsorbed, by cotransport. Sodium also crosses the tubular epithelium passively. Water is reabsorbed passively, by osmosis and diffusion, largely as a result of the movement of sodium ions. Waste organic molecules are reabsorbed to a varying extent.

7   Some substances are added to the filtrate by secretion from the tubular epithelial cells.

8   The tubular epithelial cells also have a metabolic role.

9   Renal function is assessed by measuring clearance, usually of creatinine, from the plasma.

1.4 Overview of the control of renal processes

You have seen that the major processes involved in urine formation are glomerular filtration and tubular reabsorption and secretion. It is by control of these processes that regulation of variables such as body fluids, sodium, and pH occur. First we will look at the control of glomerular filtration.

1.4.1 Control of glomerular filtration

In Section 1.3.1, you learnt that the glomerular filtration rate, or GFR, is the volume of fluid entering the tubule per unit time. The main factor influencing GFR is the pressure of the blood in the afferent and efferent arterioles. Both the nervous and endocrine systems are involved in the control of the state of dilation of the renal arterioles, and hence in the control of GFR.

What, then, is the nature of the stimulus that results in changes in the pressure in the arterioles? One important stimulus is blood pressure itself. In Section 1.3.1, we stated that GFR does not vary greatly with variations in arterial blood pressure. This is because, like other capillary systems, the glomerular system is *autoregulated*. This means that when blood pressure in the afferent arteries increases, the arterioles constrict, reducing the pressure in the capillaries. This prevents sudden large increases in pressure in the capillary bed. However, a *persistent* increase in arterial pressure will result in a general increase in pressure in the arterioles and thus also in the glomerulus, and this will cause an increase in GFR.

This brings us to an important point. The kidney actually plays a vital role in the regulation of blood pressure. This is because an increase in blood volume will cause an increase in blood pressure. Blood volume, in turn, is related to the volume of water in the body. Since the kidney is the major route by which the levels of water in the body are regulated, its normal function is essential in the regulation of blood pressure.

Now let us return to how GFR is controlled. By adjusting its own resistance to blood flow (*autoregulation*) the kidney can maintain a nearly constant GFR despite, as mentioned previously, fluctuations in arterial blood pressure. The remarkable feat is brought about by two control mechanisms: a *myogenic* mechanism and the *tubuloglomerular feedback* mechanism.

The **myogenic** (meaning *'of the muscle'*) mechanism reflects the tendency of smooth muscle of the arterioles to contract when stretched. Increasing blood pressure causes the afferent arterioles to constrict, which restricts blood flow into the glomerulus and prevents glomerular blood pressure from rising to damaging levels. Declining blood pressure causes dilation of the afferent arterioles and raises glomerular *hydrostatic* pressure (the pressure exerted by the fluid). Whereas the myogenic mechanism responds to changes in blood pressure, the **tubuloglomerular feedback mechanism** responds to the flow of filtrate within the tubule. This mechanism is controlled by specialized cells located in **juxtaglomerular apparatus**. If you look back to Figure 1.4, you will see that after the loop of Henle, the tubule passes very close to the glomerulus, in between the afferent and efferent arterioles. This arrangement is the juxtaglomerular (*'juxta'* means *next to*)

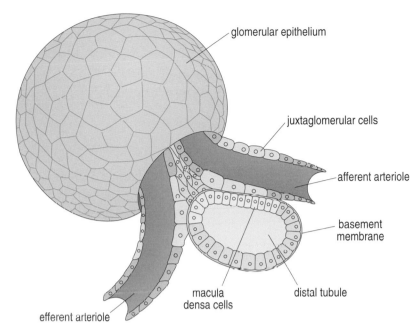

apparatus, and is shown in more detail in Figure 1.11. The specialized cells mentioned above are a special type of tubular epithelial cell known as **macula densa cells**, which lie adjacent to the arterioles, and a second type of cell, the **juxtaglomerular cells** which lie around the arterioles. These cells produce an enzyme called **renin**, which has several actions, one of which results in the production of substances that decrease the diameter of blood vessels and so increase the blood pressure. The significance of this cell arrangement is that changes in the composition and nature of the filtrate in the lumen of the tubule, such as the osmotic pressure of the filtrate, are detected by the macula densa cells which, in response, release factors that diffuse to the juxtaglomerular cells to effect changes in renin production and ultimately in the state of contraction of the arterioles, thus influencing filtration rate.

For example, if the concentration of ions/salts detected in the filtrate by the macula densa cells is high, it would indicate that the flow through the tubule is too great to allow the 'normal' amount of reabsorption in the proximal tubule and loop of Henle. This could be compensated for by a reduction in renin release, causing a reduction in the pressure in the glomerular capillaries and a resultant reduction in the GFR. Thus the flow of filtrate through the tubule would be less, and increased reabsorption would be possible, returning the levels of ions in the filtrate to 'normal'. This is illustrated as a flow diagram in Figure 1.12 (overleaf).

Under normal conditions of salt and water intake, these autoregulatory mechanisms maintain a relatively constant flow of blood through the kidneys, often over large variations in arterial blood pressure (80 to 180 mmHg; see Section 2.4.7). Consequently, large changes in water and solute excretion are prevented. However, it is important to note that these intrinsic mechanisms fail when blood pressure is extremely low, such as might result from a serious loss of

Figure 1.11 Simple diagram showing the arrangement of the specialized cells in the distal tubule and the arterioles at the juxtaglomerular apparatus.

The figure labels read: glomerular epithelium, juxtaglomerular cells, afferent arteriole, basement membrane, distal tubule, macula densa cells, efferent arteriole.

Figure 1.12 Simplified diagram showing an example of the control of GFR. An increase in sodium ion levels in the distal tubule is detected by the macula densa cells and leads to a decrease in renin secretion by the nearby juxtaglomerular cells. Reduced renin levels cause a decrease in GFR which in turn leads to increased filtrate processing and a reduction in the sodium concentration at the macula densa.

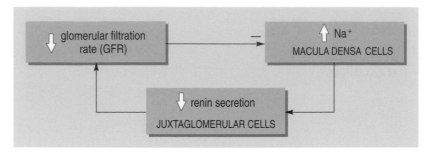

blood (haemorrhage) or when body sodium levels are low. Under these conditions autoregulation ceases. We will return to the regulation of GFR in more detail below, when considering the homeostatic regulation of water and sodium levels in the body.

1.4.2 Control of reabsorption processes

You have seen that several different processes are involved in tubular reabsorption. What may not be clear is how these processes can be controlled in order to affect the *amounts* of different molecules and ions that are reabsorbed. The answer is that there are mechanisms which allow *changes* in the membrane properties of the epithelial cells in some parts of the tubules, in particular those of the distal parts of the tubule, and also those in the collecting ducts. One of these changes allows increased reabsorption of water when water needs to be conserved.

A second feature of the anatomy of the nephron plays an essential role in the control of reabsorption, this is the long loop of Henle. You have seen that the final control of water reabsorption is in the distal part of the tubules, in the collecting ducts. The changes in the collecting ducts that allow increased reabsorption of water are only effective because the osmolarity of the extracellular fluid in this region is elevated.

● Can you work out why this is important?

● Because water is absorbed passively, by osmosis. If there was no osmotic gradient, even if the permeability of the ducts were increased, no increase in water reabsorption would occur.

As you have seen, absorption of water from the tubule is an entirely passive process, and is dependent upon the concentration of solutes in the epithelium and surrounding tissue being higher than that in the filtrate. A key factor in the control of water reabsorption, then, is the existence of a gradient of increasing osmolarity in the extracellular fluid of the medulla, as shown in Figure 1.13.

The osmolarity increases with distance into the medulla from the cortex. This gradient, to which urea makes an important contribution, is set up by a process known as the **countercurrent multiplier system** which comes into play as the filtrate passes along the long loops of Henle. The existence of the gradient means that, as the filtrate passes along the collecting ducts, the osmolarity of the extracellular fluid surrounding the ducts increases. This allows reabsorption of water, by osmosis, along the whole length of the ducts, when, under the influence

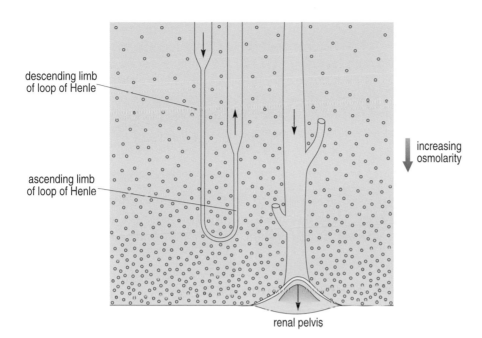

descending limb
of loop of Henle

ascending limb
of loop of Henle

increasing
osmolarity

renal pelvis

Figure 1.13 The gradient of osmolarity of the medulla (solute particles are shown as green circles).

of stimuli which you will learn about shortly, the ducts have increased permeability to water. If there were no gradient, although water could be reabsorbed in the first part of the ducts, the increase in the osmolarity of the filtrate caused by the loss of the reabsorbed water would soon prevent further water reabsorption, as the concentration of solutes in the filtrate would reach the same value as that in the extracellular fluid.

1.4.3 The glomeruli and kidney disease

Many diseases affect kidney function by damaging the glomeruli. Glomerular diseases include many conditions with a variety of genetic and environmental causes, but they generally fall into two major categories:

- *glomerulonephritis* – the inflammation of the glomeruli.
- *glomerulosclerosis* – the scarring or hardening of the blood vessels of the glomerulus and kidney.

Damage to the glomeruli reduces their ability to act as a filter, leading to proteins and in some cases red blood cells entering the filtrate and the urine (blood in the urine is called *haematuria*). Often the glomerular filtration rate is reduced, interfering with the clearance of waste products so they begin to accumulate in the blood. Furthermore, loss of blood proteins such as albumin in the urine can result in a fall in their levels in the bloodstream (the presence of proteins in the urine is called *proteinuria* (albumin in the urine is called *albuminuria*); a fall in blood protein levels is called *hypoproteinaemia*). In normal blood, albumin acts as a sponge, drawing extra fluid from the body into the bloodstream, where it remains until the kidneys remove it. When albumin is lost in the urine, the blood loses its capacity to absorb extra fluid from the body. Fluid accumulates outside of the vascular system causing the face, hands, feet and ankles to swell (a condition called *oedema*). Most forms of glomerular disease develop gradually, often causing no symptoms for many years. Not all forms of glomerular disease involve inflammation, though most do. Inflammation of the glomeruli occurs when immune complexes (interactions between antigenic material and antibodies, see this book, Chapter 4) formed in the blood or within the kidney lodge in the walls of the glomeruli. Damage can either extend to all the glomeruli or be restricted to a small region of the kidney. Figure 1.14b shows the kidney of a patient who died of ESRD as a result of chronic glomerulonephritis. Because of the damage caused to the blood vessels and tubules (affecting the supply of nutrients and clearance of waste products) the kidney is less than half the size of a normal one (Figure 1.14a); note the scarring to the cortices.

Figure 1.14 (a) Gross specimen of a normal kidney. (b) Kidney taken from a patient who died of ESRD as a result of glomerulonephritis. Note how much smaller this kidney is compared with the normal kidney shown in (a).

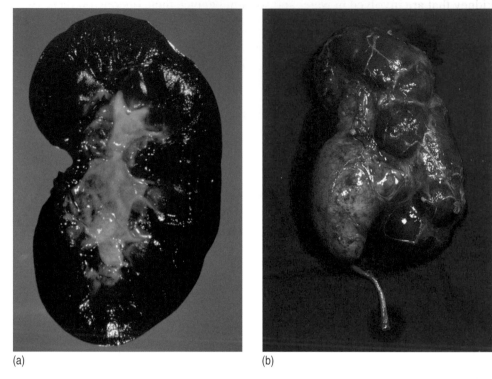

(a)   (b)

Summary of Section 1.4

1 Blood pressure in the afferent and efferent arterioles influences the rate of glomerular filtration, and the kidney plays an essential role in the regulation of blood pressure.

2 Control of GFR is made possible because of the arrangement of specialized cells at the juxtaglomerular apparatus.

3 The filtrate is monitored by the macula densa cells which in turn regulate the function of the juxtaglomerular cells.

4 The juxtaglomerular cells secrete renin, which starts a chain of events leading to a change in the diameter of the blood vessel walls, and so an alteration in the blood pressure and therefore the rate of filtration.

5 Reabsorption is controlled by mechanisms which produce a change in the membranes of tubular epithelial cells in the distal tubules and collecting ducts, and is dependent upon a gradient of increasing osmolarity in the extracellular matrix of the medulla, set up by the countercurrent multiplier system.

1.5 Regulation of body sodium and water levels

Now that you are familiar with the renal processes involved in the excretion and reabsorption of water and salts, and with the places at which these renal processes are controlled, we can turn to *how* the amounts of water and salts in the body are regulated. In this first section, we will consider the regulation of water and sodium content. Both these substances undergo only filtration and reabsorption in the kidney; they are not secreted. The control mechanisms in the kidney that are involved in water and sodium balance then, operate on these two processes.

Under 'normal' circumstances, the amounts of water and sodium taken in are the same as the amounts eliminated; there is no net loss or gain of these substances (see Table 1.3; see also Book 1, Section 3.8).

Table 1.3 Average daily intake and output of water and sodium chloride in adults.

	Water	Sodium chloride
Intake		
drunk	1200 ml	
in food	1000 ml	10.50 g
metabolically produced	350 ml	
total	2550 ml	10.50 g
Output		
loss via skin and lungs	900 ml	
sweat	50 ml	0.25 g
in faeces	100 ml	0.25 g
urine	1500 ml	10.00 g
total	2550 ml	10.50 g

This table shows only average values, and as we are all aware, the amount of water and salts taken in and excreted vary considerably according to what we eat and drink and our activities. The homeostatic mechanisms that regulate water and sodium levels are able to deal with these variations, with resultant variation in the concentration and volume of the urine produced. For example, the volume of water excreted in the urine can be between 0.4 and 25 litres per day, depending on the volume of fluid ingested. Excretion of a large volume of urine is known as **diuresis**. At this point, some of you may be thinking of the effects of drinking lots of tea, or many pints of beer; in fact alcohol has a particular effect on the control mechanisms that regulate body fluid levels, as you will see.

● What is the term used to describe a substance, such as alcohol, which increases urine output? (*Hint*: you first met this term in Book 1, Section 3.8.)

● Such substances are called *diuretics*.

- Can you think of some situations where increased volumes of water and sodium are *lost* from the body by routes other than the kidney?

- In hot weather and/or during exercise, sweating may be greatly increased, leading to an increased loss of water and sodium. During a bout of vomiting or diarrhoea, extra water and salts are lost.

The regulation of the levels of water and salts is not only important in the circumstances we have just described, but all the time, since, like everything else in the body, fluid and salt levels are in a dynamic state. You will see that the homeostasis of water and salts involves the interactions of many different systems of the body.

1.5.1 Water balance

The first step in the regulation of water levels is the detection of a change or deviation from the optimum levels of water in the body. As you know, water is the basis of all body fluids. Think, for a moment, of the fluid compartments of the body. You will remember that there are two sites in which fluid is present: inside cells (intracellular fluid), and outside cells (extracellular fluid). Extracellular fluid is present in the blood plasma, and also in the spaces between cells, where it is called interstitial fluid. As water can pass between these sites by osmosis and diffusion, a change in water levels will therefore affect the entire body. First, we will look at what happens if insufficient water is ingested.

- When the water content of the body decreases, what effect will this have on the osmolarity of the intracellular and extracellular body fluids?

- The osmolarity of the intracellular and extracellular body fluids will increase.

Any change in the osmolarity of the extracellular compartment is detected by a group of neurons in the hypothalamus, which are therefore known as **osmoreceptors.** When the osmoreceptors detect an increase in osmolarity of the interstitial fluid which surrounds them, they activate another group of hypothalamic neurons, which secrete antidiuretic hormone (ADH) (Book 2, Section 3.2).

- In which gland are the processes from the ADH-secreting hypothalamic nuclei found (look at Figure 3.10 of Book 2 if unsure of the answer) and where does ADH enter the circulation?

- The hypothalamic neurons that produce ADH extend processes to the pituitary gland (specifically to the posterior pituitary) and this is where ADH enters the blood circulation, as shown in Figure 1.15.

ADH stimulates changes in the membrane proteins of the epithelial cells of the collecting ducts in the kidney. It specifically affects a class of membrane proteins called **aquaporins** (literally '*water pores*'), proteins that form channels in the membrane that are permeable to water. Several types of aquaporin molecule have been discovered, including at least four in the membranes of epithelial cells in the kidney. ADH affects the structure and function of a specific aquaporin called AQP2, a water pore found in the luminal membrane of the epithelial cells

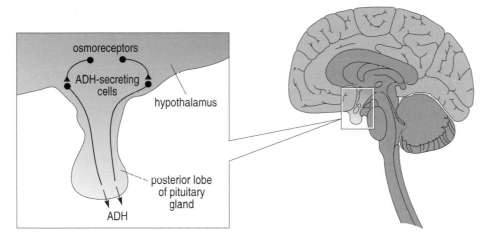

Figure 1.15 Osmoreceptors and ADH-secreting cells.

in the collecting ducts of the kidneys where urine is concentrated. AQP2 allows water from the tubule lumen to enter the epithelial cells. ADH affects the water permeability of the collecting ducts via its effect on AQP2. The acute effect of ADH, i.e. within minutes, is to increase the number of AQP2 water pores in the luminal membrane (see Figure 1.16).

● What effect do you think this might have on the passage of water through the epithelial cells?

● By increasing the number of aquaporin 'water pores' on the luminal membrane the capacity to transport water from the tubule is increased. Because more water is removed from the tubule filtrate, less is loss in the urine (hence the antidiuretic action of ADH).

Longer-term effects of ADH are to increase the transcription and translation of the AQP2 gene, resulting in a further increase in the levels of AQP2 in the epithelial cells.

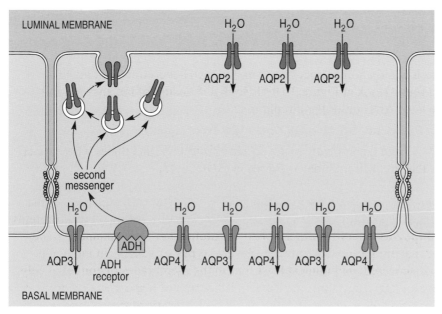

Figure 1.16 Rapid activation of AQP2 by ADH in epithelial cell of kidney collecting duct. ADH regulates the insertion of AQP2 water pores into the luminal membrane of the epithelial cells. ADH binds to a receptor on the basal membrane and activates a transduction pathway that generates second messenger molecules which promote the insertion of AQP2 proteins into the luminal membrane. Prolonged activation of the ADH receptor also increases the synthesis of AQP2 proteins, further enhancing the capacity of the luminal membrane to transport water. Since the osmolarity of the extracellular fluid surrounding the collecting ducts is high (because of the countercurrent multiplier system), water is readily reabsorbed from the tubule resulting in a concentrated urine.

Another stimulus for ADH release comes from stretch receptors in the cardiovascular system. These vascular stretch receptors detect changes in blood fluid levels. There are two types: volume receptors found at the junctions where the veins return blood to the heart, which detect small changes in blood volume, and arterial pressure receptors known as **baroreceptors** (see Section 2.4.6), which come into play when there is a severe loss of blood. Some of the vascular stretch receptors relay this information to the hypothalamus where they influence ADH secretion from nearby cells.

● You have already seen that fluid levels affect blood volume, and thus influence blood pressure. If water levels are low, what will be the effect on blood pressure?

● It will be reduced. Blood pressure is reduced if blood volume is reduced. Blood volume will be reduced if water levels in the body are low.

The decrease in blood volume detected by the vascular stretch receptors results in a decrease in their firing, which in turn causes an increase in ADH release.

The control of ADH secretion is shown in Figure 1.17.

Figure 1.17 Diagram showing the factors controlling ADH secretion.

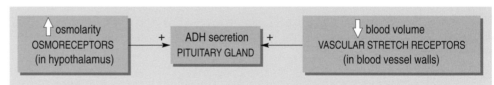

ADH, then, plays an essential role in the conservation of water at times when fluid intake is low, because it increases the reabsorption of water from the collecting ducts. As we have already stressed, however, there is another renal process which influences water levels, and that is filtration.

● You have seen that a lowering of fluid levels results in a decrease in blood pressure. How will this affect GFR, and water reabsorption?

● A general reduction in blood pressure will cause a lowering of GFR. This will cause an increased reabsorption of sodium, and hence an increase in the passive reabsorption of water.

There is another important aspect of the control of water levels which does not take place in the kidney, and that is the process of drinking itself, stimulated by feelings of thirst. Another group of cells in the hypothalamus, sometimes called *thirst receptors*, are also activated by an increase in the osmolarity of the extracellular fluid. This stimulates a feeling of thirst, which results in increased intake of water.

Since ADH plays a key role in water balance, substances that interfere with ADH production will thus cause disruption to the normal production of urine. A number of you may have experienced this. Alcohol inhibits the release of ADH. This is why consumption of alcoholic drinks seems to have a disproportionate effect on urine production, and also why, in excess, it leads to dehydration and subsequent feelings of thirst!

Diseases in which ADH production is affected also cause a disruption to the normal production of urine. Loss of the ability to synthesize ADH – for example, if the hypothalamus is damaged – is the cause of the condition known as **diabetes insipidus,** characterized by the formation of copious amounts of very dilute urine. It is usually treated successfully by the administration of ADH.

● Now that you have seen what occurs when water levels in the body are low, can you work out what will happen if water levels are high?

● ADH will not be secreted, as the osmoreceptors and baroreceptors that project to the ADH-producing neurons in the hypothalamus will not be appropriately activated. If ADH is not present, the rate of water absorption in the collecting ducts will be very low, and increased volumes of dilute urine will be produced. Of course when water levels are high, increased plasma volume occurs; this will result in an increased GFR, reducing the reabsorption of sodium. This will lead to a reduction in the passive reabsorption of water, which further contributes to an increased excretion of water.

1.5.2 Regulation of sodium levels

Sodium, like water, is not secreted in the kidney, so the excretion of sodium is dependent upon only two factors: filtration and reabsorption. Both processes are controlled in the regulation of sodium levels.

First, we will look at what happens if sodium levels are low, for example, if insufficient salt is ingested. Under these circumstances, the osmolarity of the body fluids will decrease

● What effect will a reduction of the sodium content, and thus of the osmolarity of the body fluids have on blood pressure, and how will this affect GFR?

● It will cause a reduction in blood pressure, because, when sodium levels in the body fall, the plasma volume falls. A reduction in blood pressure will cause a reduction in GFR, because the arterial pressure in the kidneys is lowered as part of the general reduction in blood pressure.

The reduction in GFR will allow increased reabsorption of sodium, and hence also of water, reducing further sodium loss. This is another example of negative feedback.

Other factors act to increase GFR when sodium levels are reduced. These involve the action of another group of hormones, and the special anatomical feature of the kidney tubule that was described earlier in this chapter. Thinking back, can you recall how the distal tubule returned to pass close to the glomerulus, and the special types of cell found there, in an arrangement known as the juxtaglomerular apparatus? This was shown in Figure 1.11.

The macula densa cells detect some aspect of reduced flow in the tubule. Although this is not yet fully understood, it is thought that this may be a lowered level of sodium ions, since more reabsorption of sodium is possible in the proximal tubule and loop of Henle if flow is reduced. Whatever the change detected, the macula densa cells respond by stimulating the juxtaglomerular cells to secrete renin into the blood. Renin does not act directly on blood vessels, but cleaves a small polypeptide, **angiotensin I**, from a precursor molecule called **angiotensinogen**, which is produced by the liver. Angiotensin I is then converted into **angiotensin II** by the action of **angiotensin converting enzyme**, which is located on the luminal (i.e. inner) surface of capillary endothelial cells, particularly those of the lung. This sequence of events, involving the three different organs, is illustrated in Figure 1.18.

Figure 1.18 Diagram illustrating how the production of angiotensin II from angiotensinogen involves the kidneys, liver and lungs.

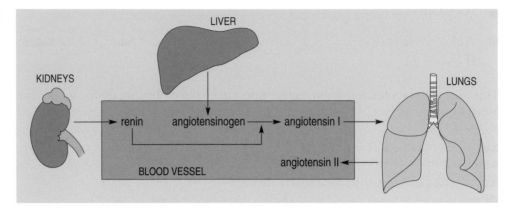

Angiotensin II is a hormone which is rapidly broken down in the circulation, but which, among other things, acts to change GFR. Angiotensin II causes a constriction of blood vessels and so an increase in blood pressure, and thus is important in the regulation of blood pressure. In the nephron, angiotensin II is particularly active on the *efferent arterioles*, thus causing an increase in GFR.

There is a third way in which GFR is increased, and that is in response to the activity of sympathetic nerves (Figure 1.19). A fall in blood pressure is detected by the volume and pressure receptors mentioned earlier and these activate

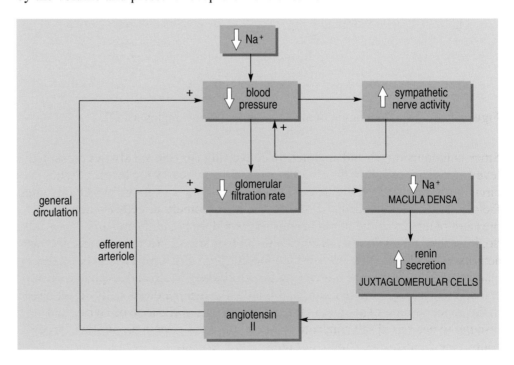

Figure 1.19 Control of GFR, shown in more detail than in Figure 1.12.

sympathetic nerves, causing a constriction of blood vessels, particularly the renal afferent arterioles which dramatically slows the rate of filtrate formation. The reduced rate of filtration then triggers the release of renin from the juxtaglomerular apparatus. The ways in which GFR is controlled are summarized in Figure 1.19.

However, the main way in which sodium levels are regulated is by alteration to the process of reabsorption. This is achieved by the action of a steroid hormone called aldosterone.

● Which organs secrete aldosterone?

● Aldosterone is produced by the adrenal glands situated at the top of the kidneys (Figure 1.1).

This hormone increases sodium reabsorption from the collecting ducts. It does this by stimulating an increase in the synthesis of sodium transport proteins in the epithelial cells in this region. (Aldosterone also stimulates the production of sodium transport proteins in the intestinal epithelium and the epithelial cells lining the ducts which empty fluid from the sweat and salivary glands; this increases sodium reabsorption and reduces sodium loss from the gut and skin.) Angiotensin II stimulates aldosterone production. This is illustrated in Figure 1.20.

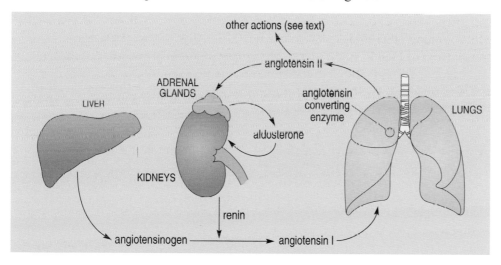

Figure 1.20 The stimulation of aldosterone release by angiotensin II.

Since angiotensinogen and angiotensin converting enzyme are always present, the levels of angiotensin II in the circulation are determined by the levels of renin in the circulation. Renin levels are determined by the kidney and are influenced by three factors. These are: sympathetic nerves; the juxtaglomerular cells (which act as pressure receptors in the blood vessels of the kidney); and the macula densa cells. So, lowering of blood pressure outside the kidney stimulates sympathetic nervous activity, which stimulates renin production; the juxtaglomerular cells also detect changes in blood pressure, this time within the kidney itself, and respond to lowering pressure by secreting more renin; and finally, the macula densa cells detect changes in the concentration of sodium in the ascending part of the loop of Henle, and respond to decreased salt concentrations by stimulating renin production. These control mechanisms are shown in Figure 1.21 (overleaf).

Figure 1.21 The control mechanisms acting to increase renin release, and hence release of aldosterone.

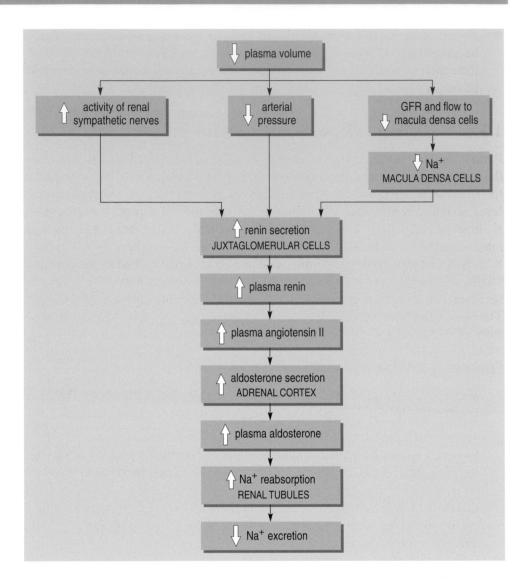

Under normal circumstances these processes regulate the levels of sodium, and thus plasma volume. This influences blood pressure, helping to maintain it at appropriate levels. Abnormally high activation of the renin–angiotensin system can contribute to high blood pressure. This can occur, for example, when a renal artery becomes partially blocked.

When sodium levels are higher than optimal, another hormone also plays a role in the regulation of sodium balance. This is **atrial natriuretic factor (ANF)** which is produced by cells in the atria of the heart. ANF inhibits sodium reabsorption in the kidneys and also increases GFR by acting on renal blood vessels. ANF secretion is stimulated by increased atrial distension, which occurs because of the increased plasma volume when sodium levels are raised.

An essential factor in the consideration of the regulation of sodium balance is the control of intake. Salt appetite is stimulated by decreased blood volume. Both thirst and salt appetite are stimulated by angiotensin II, but salt appetite, unlike thirst, is not immediate, but takes several hours to develop.

● Why will excess dietary sodium cause an increase in blood pressure?

● Excess sodium in the body fluids will cause an increased retention of water, because water will move, by osmosis, towards regions of high salt levels. Retention of water causes increased blood volume; this causes an increase in blood pressure.

1.5.3 Regulation of potassium levels

Potassium ions play an important role in all cells, and cross cell membranes to enter the cytosol by active transport, by the action of the sodium pump. It is essential that the K^+ concentration in the extracellular compartment is kept at an appropriate level, so that this pump can be effective. The regulation of appropriate extracellular K^+ levels is particularly important for the maintenance of membrane potentials in muscle and nerve cells. As with other ions, K^+ levels are regulated by the kidney: K^+ is filtered through the glomerulus, and most is then reabsorbed by the proximal tubule and the loop of Henle. In the cortical collecting ducts, however, K^+ may be secreted, and it is this process which is controlled to maintain potassium balance. The main factor affecting K^+ secretion is aldosterone, which, as you have already seen, also plays an essential role in the regulation of Na^+ levels.

Summary of Section 1.5

1   Water levels are monitored in both the intracellular and extracellular fluid compartments of the body.

2   When osmolarity of the body fluids increases, for example if water levels are low, osmoreceptors in the hypothalamus respond to the increased osmolarity of the extracellular fluid by stimulating nearby cells to secrete the hormone ADH. ADH increases the permeability of the collecting ducts to water. More water is therefore reabsorbed. This is possible because of the gradient of increasing osmolarity of the extracellular fluid around the ducts, set up by the countercurrent multiplier system.

3   Stretch receptors in the cardiovascular system detect changes in the distension of blood vessels and thus changes in blood volume. When distension is reduced, this too results in an increase in ADH secretion.

4   Thirst receptors (another type of cells in the hypothalamus) respond to increased osmolarity of the body fluids by stimulating a feeling of thirst.

5   Regulation of sodium levels is closely related to regulation of water levels. Increased sodium causes an increase in osmolarity, controlled in part by changes in water intake and excretion.

6   Sodium levels are also monitored in the filtrate as it passes through the juxtaglomerular region of the renal tubule. Macula densa cells respond to conditions in which the sodium content in the filtrate is reduced by stimulating renin production by the juxtaglomerular cells. Renin production results in the formation of angiotensin II, which acts to constrict blood vessels, increasing blood pressure, and thus GFR. In the kidney the efferent arterioles are particularly affected by angiotensin II; their constriction therefore also increases GFR. Angiotensin II also stimulates aldosterone secretion from the adrenal gland. Aldosterone acts to increase sodium reabsorption in the collecting ducts.

7   Salt appetite is stimulated by a decrease in blood volume and by angiotensin II, which also stimulates thirst.

8   Increased distension in the atria of the heart, caused by increased blood volume, causes the release of atrial natriuretic factor (ANF). ANF inhibits sodium reabsorption and acts on renal blood vessels to increase GFR, thus accelerating sodium excretion.

1.6 Regulation of pH

You have already learnt something about pH in the earlier part of this course (see Box 4.2 in Book 1). Acidity is expressed in terms of pH, and pH is directly related to the concentration of hydrogen ions (H^+), such that acidic solutions have a high concentration of H^+ and alkaline solutions have a low concentration of H^+. The pH of most body fluids is about neutral; that is, they have a pH of around 7.4. A neutral pH is 7.0, acid pH values are less than 7 and alkaline solutions have pH values greater than 7.

As with all other ions in the body, the levels of hydrogen ions must be regulated, so that potentially damaging deviations from a neutral pH do not occur. An increase in levels of hydrogen ions in the circulation is known as **acidosis** and conversely, a decreased hydrogen ion concentration is known as **alkalosis**. Hydrogen ions may be lost or gained in the body by a variety of processes; for example, they may be produced as a by-product of various metabolic reactions and can be lost during vomiting.

●   Why are hydrogen ions lost during vomiting?

●   Because of the high concentration of hydrogen ions in the stomach (Book 1, Section 4.3.4).

Most hydrogen ions in the body result because of aerobic respiration (see Chapter 3). Carbon dioxide (CO_2) combines with water to form carbonic acid (H_2CO_3), which dissociates (breaks up) to form hydrogen ions and bicarbonate ions (HCO_3^-). An equation showing the relationship of these interrelated molecules is shown below. The two-way arrows indicate that this reaction can take place in both directions; in other words, it is reversible. Furthermore, the two-way arrows indicate that these molecules establish an equilibrium; a state when the concentration for each molecule has reached a stable level.

$$CO_2 + H_2O \rightleftharpoons H_2CO_3 \rightleftharpoons HCO_3^- + H^+ \tag{1.2}$$

The important feature of an equilibrium reaction is that if it is perturbed it will readjust to re-establish its equilibrium state.

●   If we were to inhale more carbon dioxide, how would this affect the equilibrium reaction shown above?

●   An increase in the concentration of carbon dioxide will bring about an increase in the level of carbonic acid which in turn will increase the concentration of hydrogen ions.

We say the reaction will tend to move towards the right – eventually the levels of carbon dioxide and hydrogen ions will reach a point where there is no longer a net loss of carbon dioxide or a gain of hydrogen ions, the system is now in equilibrium.

Hydrogen ions can bind, in a reversible manner, to bicarbonate ions and also to other molecules, including proteins such as haemoglobin. This has the effect of buffering (neutralizing) the hydrogen ions. The regulation of pH in the body then, occurs by buffering of hydrogen ions, until the excess can be excreted by the kidney.

The regulation of the levels of circulating hydrogen ions that occurs in the kidney is actually achieved by changing the levels of bicarbonate ions in the plasma.

● What would you expect to happen to the free hydrogen ion concentration in the plasma if bicarbonate ions are removed and excreted in the urine?

● Removal of bicarbonate ions would have the effect of 'pulling' the reaction shown above to the right, thus raising the levels of free hydrogen ions in the circulation. Conversely, if extra bicarbonate ions are added to the plasma, it has the effect of buffering an equal number of hydrogen ions in the circulation, reducing the concentration of free hydrogen ions.

How, then, do these processes occur in the kidney? Bicarbonate ions are freely filtered through the glomerulus. Tubular epithelial cells produce hydrogen ions and bicarbonate ions. The hydrogen ions are secreted into the lumen, the bicarbonate ions move, down the electrochemical gradient produced by the active transport of sodium ions, into the circulation. This is shown in Figure 1.22. The bicarbonate ions that have been filtered into the lumen of the tubules then combine with the secreted hydrogen ions to form carbon dioxide and water (according to the equation above) which are reabsorbed into the epithelial cells. Thus, although the bicarbonate ions have not been reabsorbed as such, there is no net loss of bicarbonate ions from the circulation.

How does this process result in regulation of hydrogen ion levels in the circulation? Part of the answer is that when all the filtered bicarbonate is bound to secreted hydrogen ions, any extra hydrogen ions that are secreted bind to other, *non* bicarbonate buffers in the filtrate. When this happens, there is a net *loss* of hydrogen ions, and a net *gain* of bicarbonate ions, from the plasma.

Another way in which bicarbonate levels in the plasma can be increased is by the production of bicarbonate in epithelial cells of the proximal tubule, by a chemical reaction involving the breakdown of the amino acid glutamine. This was one of the processes referred to earlier in this chapter when discussing tubular metabolism. The resulting bicarbonate enters the circulation, while ammonium ions are excreted. This is shown in Figure 1.23 (overleaf). Plasma pH is hence regulated by control of these processes, but it is too complex to concern us here.

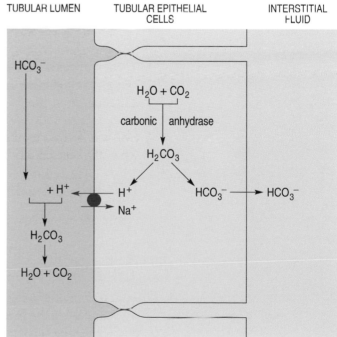

Figure 1.22 Secretion of H^+ ions by tubular epithelial cells.

Figure 1.23 Metabolism of glutamine by tubular epithelial cells produces bicarbonate ions.

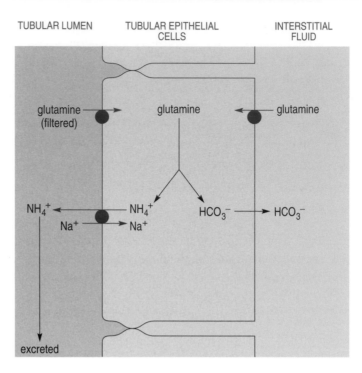

The respiratory system also plays an important role in pH regulation, as you will see in Chapter 3.

Summary of Section 1.6

1   The pH of body fluids is regulated by buffering of hydrogen ions and by secretion of hydrogen ions by the renal tubules.

2   Buffering of hydrogen ions in the circulation is mainly by bicarbonate ions, but also by other molecules such as proteins.

3   In the kidney, secreted hydrogen ions bind to filtered bicarbonate ions, excess hydrogen ions then bind to other components of the filtrate. Since the secretion of hydrogen ions by the tubular epithelial cells is accompanied by the production of an equal number of bicarbonate ions which return to the circulation, there is no net loss of bicarbonate.

4   Additional bicarbonate can enter the circulation as a result of the metabolism of glutamine by the tubular epithelium.

1.7 Movement of urine to the bladder and micturition

After passage through the tubules, urine is moved along the ureters to the bladder. The ureters are tubular structures, which are composed of smooth muscle and have an inner lining of epithelial cells. Movement of the urine is by a peristaltic type of action of the smooth muscle.

●   What is peristaltic movement? (You met this term in Book 1.)

●   Peristaltic movement or peristalsis is a progressive wave-like contraction of the muscle wall that moves substances along hollow structures or organs.

The bladder is able to store quite large volumes of urine (300–400 ml), the smooth muscle that forms the major part of the bladder wall relaxing to accommodate the increasing volume of fluid. Under normal circumstances, urine does not leak out of the bladder, because of two specialized areas of muscle, the **internal urethral sphincter**, which is composed of smooth muscle, and the **external urethral sphincter** which consists of skeletal muscle (Figure 1.24).

● Where have you come across the term sphincter before?

● You met this term in Book 1, Section 4.2.1, when discussing the sphincters that regulate the flow and composition of the gut contents – the pyloric sphincter of the stomach and the sphincter of Oddi in the pancreatic duct.

The anatomical arrangements of the muscles are such that the internal sphincter remains closed when the muscle of the main body of the bladder is relaxed. Both the bladder smooth muscle and internal sphincter muscles are under involuntary control, by neurons of the autonomic nervous system. The external sphincter, on the other hand, in most adults at least, is controlled voluntarily.

Emptying of the bladder is known as **micturition**. As the bladder fills, stretch receptors in the muscle are stimulated. These cause a contraction of the main bladder muscle and relaxation of the internal sphincter. Neuronal processing of this response occurs in the spinal cord, but signals are also relayed to the brain, giving the sensation of fullness. In adults, as the bladder fills, contraction of the bladder muscle is stimulated. Once a certain volume is reached (300–400 ml), the contraction is strong enough to begin to open the internal sphincter. As this occurs, the nerves originating in the spinal cord that normally keep the external muscle contracted are inhibited, resulting in relaxation of the external sphincter; however, this final step is under a degree of voluntary control from descending pathways from the brain. Once both sphincters are open, the contraction of the bladder muscle forces urine out of the urethra. The pathways involved in these processes are summarized in Figure 1.24.

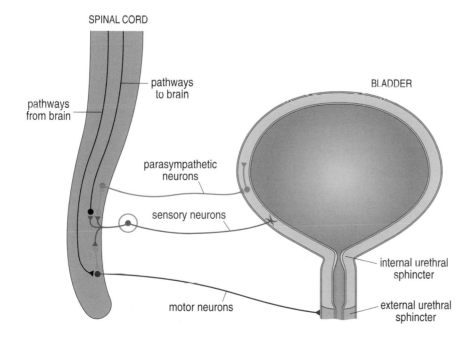

Figure 1.24 Diagram showing the nerve pathways involved in micturition. (The green circle denotes the dorsal root ganglion.)

During childhood, voluntary control of the external sphincter is learnt, by activation of pathways descending from the brain. These can cause initiation or prevention of urination, at will. If the spinal cord is injured, the ascending and descending pathways can be damaged, leading to loss of voluntary control of urination. Childbirth involves stretching of the pelvic floor muscles and can cause a physical trauma of tissues that lie close to the birth canal. This can also lead to a loss of contractility of the external sphincter and surrounding muscles, leading to *stress incontinence*, but exercising the muscles of the pelvic floor can rectify this. (It should be noted that stress *here* refers to *physical* stress or damage to tissue and not to the physiological and psychological stress covered in Book 4.)

Men can develop difficulties in micturition as they age, usually due to an enlargement of the prostate gland (Figure 1.25), as illustrated in Case Report 1.1. The prostate is a walnut-sized gland that forms part of the male reproductive system and is composed of two lobes, or regions, enclosed by an outer layer of tissue. The prostate is located in front of the rectum and just below the bladder. Importantly, it also surrounds the urethra. Whilst not all of the functions of the prostate are fully understood, it is involved in the process of ejaculation by introducing, into the urethra, fluid which forms part of the ejaculate; this fluid energizes the sperm and reduces the acidity of the vaginal canal.

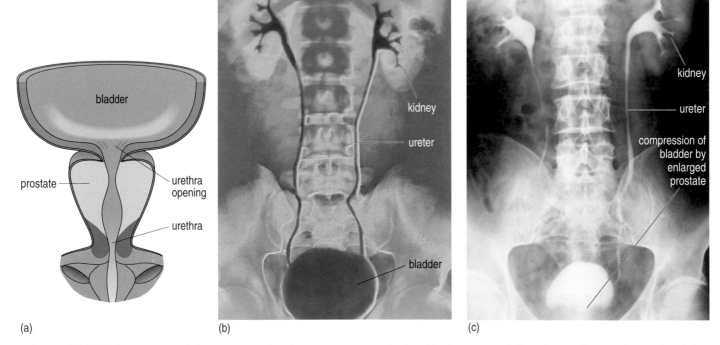

(a) (b) (c)

Figure 1.25 Enlargement of the prostate gland occurs commonly in elderly men and the obstruction to the neck of the bladder impairs urination. (a) Diagram showing the prostate gland in relation to the bladder and urethra. (b) An intravenous urogram (IVU) showing the bladder and urinary tract of a normal individual. An IVU is an X-ray image in which the urinary tract is made visible by the intravenous injection of a radio-opaque contrast medium which becomes concentrated in the urine. (c) An IVU taken from a patient with an enlarged prostate. Compare the bladder shape with that of the normal individual shown in (b). The raised, uneven outline of the base of the bladder in (c) indicates an abnormal prostate.

Case Report 1.1   Prostate enlargement

John is 69 years old and has been married to Gwen for 42 years. They have two children, both of whom are married with children of their own. Their daughter, Sarah, lives locally, and John and Gwen help to look after their grandchildren several days each week, while Sarah and her husband are at work. Their son, Stephen, is living with his family in the Netherlands. John retired from his job as an engineer four years ago and he has generally been in good health. However, over the last two years he has been increasingly troubled by urinary symptoms. He needed to get up in the night to pass urine more often – often four or five times. After occasional evenings in the village pub with a few pints of beer, he found he was getting up at night as many as eight or nine times, and felt exhausted the next day. He was also troubled by frequency during the day, needing to go to the toilet half hourly sometimes, and found that he sometimes dribbled after passing urine. He tried cutting down on his fluid intake but this did not help the problem. He felt more and more worried about going out, particularly to unfamiliar places as he was worried that he would not be able to find public toilets as often as he needed to. He also became anxious about long journeys particularly if they might entail motorway driving, in case there was a hold-up and he wouldn't be able to get to the toilet. His son and family invited him and Gwen to visit them in the Netherlands but John told Gwen that he was too worried about the journey. Gwen frequently told him to go to the GP but John kept putting it off. One day Gwen, exasperated, phoned and made John an appointment and he agreed to go.

John's GP listened to his symptoms carefully, and explained that it sounded as if John had a problem with his prostate which, he said, was common in men of his age. He then asked him if he could provide a urine sample. He carried out a urinalysis and told John that the result was normal, indicating no infection. He also checked whether John had had any symptoms of a urinary tract infection (such as pain on passing urine) but John denied this. He then explained that he would need to examine him by putting a finger up into his 'back passage' (i.e. rectum) to feel his prostate. After the examination he told John that his prostate was enlarged, and that this could be a benign

(non-cancerous) enlargement, called 'benign prostatic hypertrophy' (BPH), but there was a small possibility that it could be cancerous so he should have further tests (it should be noted that the incidence of prostate cancer is relatively common). He explained that the next stage was to do a blood test for the prostate-specific antigen (PSA) which would give an idea as to whether the enlargement of John's prostate was cancerous. He gave John a leaflet explaining the test and what it meant and gave John the option of coming back for the blood test or having it now. John said he would prefer the blood test to be done straightaway, which his GP did. The GP asked him to make an appointment in two weeks time so he could discuss the test result with him, and he gave John some other written information about prostate conditions.

John was naturally concerned but was encouraged, after reading the information, that there were various options for him depending on his diagnosis. He and Gwen decided not to tell the rest of the family yet as they didn't want to worry them. Two weeks later, his GP informed him that his PSA was slightly raised to 8 nanogrammes (ng) per ml (the normal value for his age being under 4.5). The GP explained that there was a slight possibility that John's prostate problem was caused by cancer and said that he would send a referral form to the hospital by fax and that John would be contacted by phone to make an appointment. Although John was worried about his PSA result, he was hopeful that as the PSA was only slightly raised, his problem might yet turn out not to be cancer. Two days later John received a phone call from the urology cancer specialist nurse who explained that John would need to have an ultrasound guided biopsy, which would be done in the X-ray department. He talked to John about what this entailed and John decided to have this carried out the following week. The nurse explained that the test involved an ultrasound probe being put into his rectum and measurements of the prostate made, and then samples of tissue would be extracted from different areas of his prostate gland, with a special needle, and examined in the laboratory. To prevent any infection (from contamination of the prostate by the probe passing through the rectum) the nurse said that he would send him antibiotics (metronidazole and ciprofloxacin) to take before and after the test, and also advised that he should take 1 g of paracetamol about an hour before the biopsy, and try to have his bowels open before attending to make the procedure more comfortable. The nurse said he would send John some leaflets; these would include websites and phone numbers to ring for further information. He assured John that he could be contacted at any time for more information or to answer questions.

Gwen accompanied John to the hospital, where they met the urology cancer specialist nurse. He explored with them what they understood about the tests carried out so far, and checked whether they had any more questions before the biopsy was performed. John found the procedure uncomfortable but bearable. Afterwards, he was given an appointment to attend outpatients two weeks later, to see the urology consultant and the nurse so that his test results could be discussed. John and Gwen arrived for the appointment and were very pleased when the consultant informed them that no cancer had been found. He explained that the problem was probably therefore benign prostatic hypertrophy (BPH) and that he would ask John's GP to perform a repeat PSA in six months' time. In the meantime, he suggested that John should take medication, a drug called doxazosin, to help 'firm up' the prostate gland, and he also advised John to have an ultrasound of his abdomen to look at his prostate and bladder, and check his urine flow rate. John was very relieved that no cancer had been found. He asked the consultant about surgery for his enlarged prostate as he had read about it in one of the leaflets. The consultant said that this was sometimes necessary and a transurethral prostatectomy was the usual procedure, which involved the prostate being operated on through the urethra. He suggested that John come to see him again after his repeat PSA, his ultrasound, and after trying the medication. He explained that there was other medication that could be used to shrink the prostate.

If the prostate enlarges, the layer of tissue surrounding it prevents it from expanding outwards, instead causing it to press inwards, against the urethra like a clamp on a garden hose. Over time, the bladder wall becomes thicker and irritable. The bladder begins to contract even when it contains small amounts of urine. Eventually, the bladder weakens and loses the ability to empty itself.

John was greatly reassured that he had BPH and not prostatic cancer, a condition which can be life threatening. The incidence of prostatic cancer in the developed world appears to be increasing and for reasons which are not fully understood, certain ethnic groups seem to be at more risk than others. Figures from the United States for 1995 indicate that black Americans are more susceptible to this disease than white Americans, with 200 black men diagnosed per 100 000 of the population compared with 124 for white men. A similar racial difference is reflected in the mortality rates, with 54 black men per 100 000 of the population dying from prostatic cancer compared with 23 per 100 000 for white men (Stanford et al., 1999). The reason for John's concern is that the symptoms of prostatic cancer are not dissimilar to those for BPH, in that there is often a difficulty to start or sustain a flow of urine coupled with the need to urinate frequently, often at night. Other symptoms include difficulties in having an erection, blood in the urine or semen and frequent pain in the lower back, hips or upper thigh. In the UK more than 20 000 men are diagnosed with and 9500 die of prostatic cancer each year (Nursing Standard, 2002).

1.8 Endocrine role of the kidneys

In addition to their major role in the regulation of levels of body fluids, salts and pH, the kidneys also have an endocrine role. The hormone erythropoietin is actually produced in and released from the kidney itself (although it is also produced by the liver).

●   What is the role of erythropoietin? (See Section 3.3.1 of Book 2 if unsure.)

●   Erythropoietin stimulates the proliferation of undifferentiated precursor cells present in the bone marrow. These precursors give rise to red blood cells.

Erythropoietin is released from capillary endothelial cells in the kidney; its release is stimulated by a decrease in oxygen concentration in the blood supply to the kidneys.

The kidneys also play an important role in the hormonal control of calcium homeostasis. About 50% of the plasma calcium is filtered at the glomerulus; the rest of the calcium is bound to proteins which are too large to be filtered. After filtration, calcium is reabsorbed in the tubules; it is not secreted. The levels of calcium in the plasma are, in part, regulated by control of the reabsorption process.

Two hormones play a major role in the regulation of plasma calcium levels; these are **parathyroid hormone**, and the active form of vitamin D. Parathyroid hormone is produced by the parathyroid glands, which lie in the larger thyroid gland (Book 2, Chapter 3). When plasma calcium levels are low, parathyroid hormone secretion is stimulated.

Parathyroid hormone has several actions. In addition to increasing the reabsorption of calcium by the tubular epithelial cells, it increases calcium release from bone and activates vitamin D (which results in increased absorption of calcium from the gut). Thus plasma calcium levels are regulated, as shown in Figure 1.26.

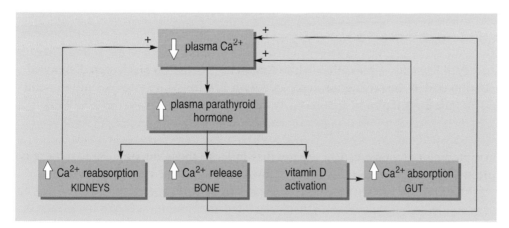

Figure 1.26 Regulation of plasma calcium levels.

1.9 Diuretics and kidney disease

A number of common conditions are treated by drugs that affect kidney function. One of the most commonly prescribed types of drug is **diuretics**. These act to increase the volume of urine excreted, and do this by inhibiting sodium reabsorption.

● Can you explain how inhibition of sodium reabsorption results in an increase in the volume of urine excreted?

● Water is reabsorbed passively, by osmosis, largely as a result of sodium reabsorption, which raises the osmolarity of the extracellular fluid. Inhibition of this process results in more urine being excreted.

Diuretics are used to treat conditions in which sodium balance is disrupted because of a failure of the kidneys to excrete sodium normally. Such a failure results in an increase in water retention in the extracellular spaces.

● What symptoms does this condition give rise to?

● This condition is known as *oedema* and is, for example, a cause of swollen ankles (see Section 1.4).

A common cause for the failure of the kidneys in this way is *congestive heart failure*, in which the contractility of the heart is reduced, resulting in a reduction of blood pressure, which, in turn, causes a reduction in GFR, reduced flow in the tubules and hence, via the renin–angiotensin system, increased aldosterone production and sodium reabsorption.

Diuretics are also used to treat high blood pressure, or hypertension. Treatment results in lowering of the content of water and sodium in the body, and reduction in arteriolar dilation.

As mentioned in the introduction to this chapter, diseases affecting the kidney are all too common. The treatment for these conditions is **dialysis**, which acts essentially as an artificial kidney system (Figure 1.27). The process involves diverting the blood from an artery into semipermeable tubing. That is, it allows the passage of water, small molecules (e.g. urea) and ions but prevents the movement of large molecules (e.g. proteins). The tubing is immersed in a large volume of a salt solution with a composition similar to plasma, but without proteins. This is called the *dialysing fluid*. The composition of this fluid is crucial. It does *not* contain substances which are normally excreted at the kidney, and which build up in the plasma of individuals with kidney failure (e.g. urea, sulfate and phosphate ions). These substances therefore diffuse *out* of the blood into the dialysing fluid. The fluid contains *higher* levels of glucose and bicarbonate ions than those typical of a healthy individual. These substances then, *enter* the blood from the dialysing fluid. Ions that do not build up in the plasma of people with kidney disease (e.g. Ca^{2+}, Mg^+), and are at similar levels in the plasma and dialysing fluid simply exchange across the tubing; their levels in the blood are not altered much by dialysis. After dialysis, the blood is returned to the body, via a vein. This form of dialysis is referred to as *haemodialysis*.

An alternative approach is not to remove the blood from the body and use an artificial membrane to cleanse it, but instead, to introduce the dialysing fluid into a body cavity, in this case the abdomen, allowing the blood to be 'cleaned' inside the body. This is achieved because the organs of the abdomen are covered by a thin membrane (tissue layer) called the *peritoneum* which allows waste products to pass through it and is very rich in small blood vessels. By passing dialysing fluid into the peritoneal cavity it is possible to filter and remove waste from the blood. This approach is referred to as *peritoneal dialysis*.

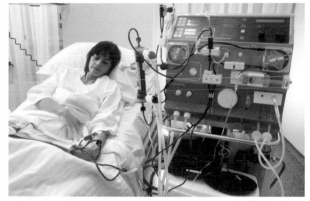

Figure 1.27 A patient undergoing haemodialysis

Summary of Sections 1.7–1.9

1   Urine passes from the kidney to the bladder via the ureters. Urine is held in the bladder by the action of two specialized areas of muscle, the internal urethral sphincter and the external urethral sphincter. Micturition (or urination) is stimulated by distension of the smooth muscle of the bladder wall. This activates stretch receptors, which trigger a local reflex contraction of the bladder muscle and relaxation of the internal sphincter, and also a sensation of fullness, which leads to voluntary relaxation of the external urethral sphincter.

2   The kidneys have endocrine functions. The hormone erythropoietin, which stimulates the proliferation of red blood cell precursors in the bone marrow, is produced in the kidneys.

3   The kidneys also play an important role in the control of calcium homeostasis, by activating vitamin D, which results in increased calcium absorption from the gut. Parathyroid hormone (produced in the parathyroid glands) stimulates this activation in the kidney, and also increases calcium reabsorption in the kidney tubules.

4   A number of common conditions, such as high blood pressure, are treated by substances called diuretics. Diuretics increase the amount of urine excreted, by inhibiting sodium reabsorption.

5   Individuals with kidney disease undergo dialysis, during which wastes are removed from the blood or via the peritoneum.

1.10 Conclusion

In this chapter we have focused primarily on the role of the urinary system in homeostasis. You have seen that the kidney plays an essential part in maintaining the balance of fluids and ions in the body, and in the excretion of waste molecules. However, we hope that it has also become clear that the kidney does not function in isolation, or simply under the influence of the nervous and endocrine systems. Normal kidney function, and hence the regulation of the amounts of water and other essential substances in the body involves *interaction* with other organ systems, some of which might initially seem to be unlikely partners with the urinary system, such as the lungs. Another important point that should be stressed is the *dynamic* nature of the processes that have been described in this chapter, which are continually going on in our bodies.

Questions for Chapter 1

Question 1.1 (LOs 1.1 and 1.2)

Substance X is present at a plasma concentration of 8 mg per ml and is excreted at a rate of 2 mg per min in the urine. Outline the possible fates of X in the kidney. What is the clearance of X?

Question 1.2 (LO 1.2)

List the fates of (a) glucose, (b) sodium and (c) water in the kidney tubules.

Question 1.3 (LO 1.3)

Complete Figure 1.28 to make a flow diagram. Show with arrows what will happen to the variables indicated after a period of severe sweating (draw arrows between appropriate variables). Which of the indicated variables must be confined within a narrow range (i.e. which are regulated variables)?

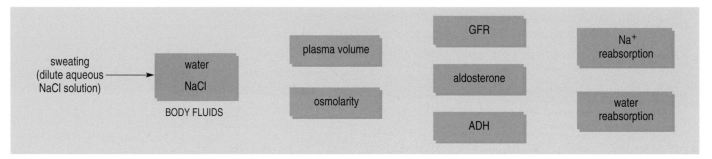

Figure 1.28 For use with Question 1.3.

Question 1.4 (LO 1.4)

Severe diarrhoea results not only in the loss of water, but also of large amounts of bicarbonate ions. What effect would this have on the pH of the body fluids, and how would a balance be restored?

Question 1.5 (LO 1.5)

What effect does decreased oxygen concentration in the circulation have on kidney function?

References

Nursing Standard (2002) (online) *Patient Information; Prostate Cancer*. Available from: http://www.nursing-standard.co.uk/archives/ns/residentpdfs/patientcards/9prostatecancer.pdf [Accessed January 2005].

Stanford, J. L., Stephenson, R. A., Coyle, L. M., Cerhan, J., Correa, R., Eley, J. W., Gilliland, F., Hankey, B., Kolonel, L. N., Kosary, C., Ross, R., Severson, R., West, D. (1999) *Prostate Cancer Trends 1973-1995*. Bethesda: SEER Program, National Cancer Institute.

CHAPTER 2

CIRCULATION

Learning Outcomes

After completing this chapter, you should be able to:

2.1 Describe and illustrate the basic anatomical components of the cardiovascular system.

2.2 Explain how heart rate is regulated by neural activity.

2.3 Explain the fundamental physiological principles of how the heart pumps blood and describe the path of electrical stimulation that leads to contraction of the ventricles.

2.4 Distinguish normal and abnormal electrocardiograms.

2.5 Outline the factors, and their mathematical interrelationships, affecting cardiac output, blood flow and blood pressure in the cardiovascular system.

2.6 Outline the main diseases and abnormal conditions of the cardiovascular system and define the action of specific drugs prescribed to treat them.

2.1 Introduction – a pumping heart and breathing lungs

Look after your heart and it will look after you.

In Chapters 2 and 3 of this book, we discuss the two systems in the body that seem to epitomize the very essence of life – the **circulatory system** and the **respiratory system**. Most of us take the seemingly effortless functioning of these vital systems for granted. However, without a pumping heart (the pump) and breathing lungs (the bellows), either real or artificial, life would stop. The heart of a person aged 72 years will have beaten roughly 2500 million times and they will have taken approximately 500 million breaths over their lifetime – simply amazing! What piece of machinery can boast such an intense and varied working history? In these chapters, you will learn about the anatomical structures and physiological functions of the circulatory and respiratory systems and how they are intimately associated and integrated. We will also learn how a wide range of homeostatic mechanisms enable these two fundamental systems to respond and adapt throughout life to biological and physical events occurring inside and outside the body – both at rest and during exercise, as well as in health and in disease.

2.2 Why do we need circulatory and respiratory systems?

Take a moment to consider your answer to the question posed above…and then read on.

The physician William Harvey (1578–1657) was the first person to identify the correct function of the heart, lungs, blood vessels and blood in the body. In his book entitled *Exercitatio Anatomica de Motu Cordis et Sanguinis in Animalibus* ('An Anatomical Treatise on the Movement of the Heart and Blood in Animals'), Harvey proposed that we need '…a heart to act as a pump that repeatedly moves a small volume of blood forward through a closed system of blood vessels which form a circular path that eventually drains back into the pump' – the circulatory system. He then correctly proposed that 'blood travels to the lungs to mix with the air' – the respiratory system. His theory superseded the centuries-old idea of Aristotle (382–322 BC), which suggested that air travels directly to the heart to 'mix' with the blood, thus producing a 'vital force' (called '*pneuma zotikon*' or 'life spirit') that is subsequently transferred to other organs through the blood vessels.

We now know that every living cell in the human body needs a continuous supply of oxygen (O_2) to simultaneously fuel a multitude of diverse metabolic processes. Oxygen is indeed a crucial gas. It combines with the carbon and hydrogen atoms in food material (usually glucose) and this process provides energy for biochemical reactions in cells that allow work such as muscular activity, cell division, digestion or brain activity to be carried out. The production of energy results in carbon dioxide (CO_2), water (H_2O) and heat as by-products. Metabolic waste products, such as CO_2, would be toxic if allowed to accumulate, so they must be removed. The combined respiratory and circulatory systems ensure the efficient transfer of O_2 'into' the body and of CO_2 'out' of the body.

The circulatory system provides an internal transport system, transporting not only respiratory gases but also all the chemicals needed by the cells for them to function correctly, and it also transports their waste products. Even heat can be transported by the blood; for example, during the process of body temperature regulation, the blood takes heat away from regions of production (e.g. exercising muscles) to regions of lower temperature, such as the skin, where it can be lost via radiation to the external environment.

● Can you think of any other important uses of an internal transport system?

● One of the critical functions of the circulatory system is to transport white cells and antibodies (the primary 'defenders of our health'), which are produced by the body's immune system and attack invading organisms, when and wherever they have breached the body's defences (see Chapter 4 for more details).

The circulatory system can be divided into two parts, the **cardiovascular system** and the lymphatic system (you will have come across the lymphatic system in Book 1, Section 1.6). These two parts work together in harmony – that is, they are functionally integrated, thereby providing a continuous movement of fluid around the body which reaches all living cells. This allows the circulatory system to carry substances produced in one part of the body to another part; for example, nutrients are transported from the gut to the rest of the body, and hormones are carried from endocrine glands to their target tissues.

From this brief introduction, we can see that the circulatory and respiratory systems are essential to the proper balanced functioning of the body. In order for these two systems to function optimally, they have to be accurately controlled – this ongoing control is critical and fundamental for a healthy body. Without appropriate homeostatic control, each system will start to function abnormally – for example, irregular heart rhythms or rapid breathing due to chemical imbalances in the blood. If unchecked, these conditions may seriously affect the normal balanced operation of the circulatory and respiratory systems to such an extent that their functioning within healthy limits may be severely compromised.

2.3 The cardiovascular system – fluid, pump and pipes

The cardiovascular system is a closed system consisting of a fluid (blood), a pumping device (the heart), and a system of pipes (the vasculature). Each of these components is considered in turn.

2.3.1 Blood – the vital fluid

Blood is a fluid comprising red cells (**erythrocytes**), several types of white cell (**leukocytes**) and **platelets**, all suspended in a yellowish, watery liquid called plasma. The components of blood and their diverse functions are summarized in Table 2.1 (overleaf).

Erythrocytes were originally termed red cells because they contain the respiratory pigment haemoglobin, which turns red when it combines with oxygen (we shall learn more about this important pigment in Section 3.2.4). All erythrocytes are structurally homogeneous and do not possess cell nuclei. They are disc-shaped with centres that are highly indented on both surfaces – seen from the side, their shape resembles a figure-of-eight (e.g. ∞). This shape is called 'biconcave', which is similar to the shape of the correcting lens for myopia shown in Figure 2.11a in Book 2. Erythrocyte membranes are highly flexible, a structural property that allows them to deform and squeeze through even the smallest of blood vessels in the body, and on occasions, even through vessel walls (see Section 2.3.3). In contrast, leukocytes lack the respiratory pigment and therefore appear white. These white cells can be divided into several discrete subtypes each with a different structure, function and number (Table 2.1).

Mature blood cells have a comparatively short lifespan (about 120 days for erythrocytes) and are renewed continuously by a process called **haematopoiesis**, a process initiated by the hormone erythropoetin, which stimulates the proliferation of the undifferentiated precursor cells in bone marrow. Although blood contains many different functional types of cells (Table 2.1), they are all produced from this common population of precursor stem cells in the haematopoietic tissue of bone marrow (Book 2, Section 3.3.1) found mainly in the pelvis, sternum, vertebrae and skull. Each erythrocyte travels about

Table 2.1   The composition of blood and the functions of its constituents. *Note*: this table contains several terms that will be unfamiliar at present; they will be explained later in this book.

Constituent	Description and functions
(a) Cellular composition	
Erythrocytes; 5 000 000 per cubic millimetre (mm^3).	Erythrocytes are small circular biconcave discs (approx. 8 μm* in diameter); transport O_2 and CO_2. Produced in the bone marrow.
White cells, also called leukocytes; 7000 per mm^3. Comprise granulocytes, monocytes and lymphocytes.	Granulocytes (approx. 16 μm* in diameter): – *neutrophils*, cells that engulf bacteria and debris; – *eosinophils*, important in allergic reactions; – *basophils*, release histamine and heparin, which are important in allergic reactions. Monocytes:– transported cells that become tissue macrophages. lymphocytes: – *B lymphocytes*, produce antibodies; – *T lymphocytes*, involved in cell-mediated immune responses.
Platelets; 250 000 per mm^3.	Involved in the process of stopping blood loss from a damaged blood vessel (haemorrhage).
(b) Plasma composition	
Water	90–95% of plasma is water, which maintains the normal hydration of the body and provides a medium for both intra- and extracellular reactions, as well as acting as a transport solvent for many essential ions. Involved in the transfer of heat around the body.
Plasma proteins	*Albumins*; constitute about 60% of all plasma proteins and regulate passage of water and diffusible solids through the capillary walls. *Fibrinogen*; constitutes about 4% of all plasma proteins and is essential for blood clotting (see also Book 1, Section 2.10). *Globulins*; constitute about 36% of plasma proteins. They are divided into alpha-, beta- and gamma-globulins: alpha- and beta-globulins transport lipids and fat-soluble vitamins in the blood; gamma-globulins are antibodies.
Plasma ions	Sodium (Na^+), chloride (Cl^-), potassium (K^+), calcium (Ca^{2+}), iodide (I^-), magnesium (Mg^{2+}) and phosphate ions, sometimes represented as 'P_i'.
Nutrients	Glucose; the body's most available source of energy. Amino acids; required for protein synthesis. Lipids; components of cell membranes and can be used as a fuel store.
Hormones	Steroid hormones; e.g. sex hormones. Peptide hormones; e.g. insulin and antidiuretic hormone (ADH). Amines; e.g. adrenalin.
Waste products	Metabolic waste products, including lactate and nitrogenous waste from protein metabolism (urea) and from nucleic acids (uric acid) – see Table 1.1.
Gases	The main gases dissolved in plasma are oxygen, nitrogen and carbon dioxide: oxygen is transported by erythrocytes with a small amount dissolved in the plasma; nitrogen is carried exclusively in the plasma; carbon dioxide is carried by erythrocytes and plasma, both in solution and as bicarbonate ions (HCO_3^-).

* 1 μm = one millionth of a metre = 0.000 001 m.

750 miles around the vasculature during its lifetime. Most erythrocytes end their mortal days in the spleen. As elderly erythrocytes try to squeeze through the narrow winding capillary network of the spleen, their fragile membranes can easily rupture and the cells die.

In adult humans, blood accounts for approximately 7–9% of the body weight, so for a person weighing 70 kg, some 5.6 litres of blood are circulating around the body. However, for a newborn baby weighing 3.2 kg (7 lb), blood volume is only 0.25 litres (250 ml), which is less than 5% of the adult value – an important consideration when transfusing blood to a small baby!

As can be seen in Table 2.1, leukocytes play a strategic role in the body's immunological defence mechanism; they are the mobile units of the body's continual fight against infection and disease and will be discussed in detail in Chapter 4.

Platelets are involved in maintaining the integrity of the wall of blood vessels (the **vascular endothelium**) and have a key role in the control of bleeding (haemostasis); see Section 2.6.5.

Erythrocytes greatly outnumber leukocytes by 750 : 1. Red and white cells, as well as platelets, can be separated from the plasma by centrifuging (spinning at high speed) a blood sample in a small test-tube. By measuring the height of the column of compacted erythrocytes relative to the total height of the column of blood, the volume percentage occupied by the packed erythrocytes can be determined – this is called the **haematocrit ratio**. The normal haematocrit ratio is about 45%.

- Given that erythrocytes in the blood carry oxygen and that the oxygen concentration of the atmosphere decreases with altitude, would a person living at high altitude and one living at sea level have different haematocrit ratios? ...And why?

- In order to maintain normal levels of oxygen in their blood, a Peruvian living high up in the Andes would have more erythrocytes per unit volume of blood compared with a resident of New York; consequently their haematocrit ratio would be higher.

(a) Major disorders of the blood

A common blood disorder is anaemia. Anaemia refers to a reduction in the capacity of erythrocytes to carry oxygen and is characterized by a low haematocrit ratio of about 30%. Four important examples of anaemia are:

(i) *nutritional anaemia*, such as *iron deficiency anaemia*, where erythrocytes have a low haemoglobin content and thus a reduced ability to transport oxygen;

(ii) *pernicious anaemia*, where insufficient amounts of vitamin B_{12} (see Book 1, Sections 3.7.2 and 4.3.6) are absorbed from the gut, leading to a reduced production and maturation of erythrocytes;

(iii) *haemorrhagic anaemia*, which can result from severe blood loss through a bleeding wound or excessive menstrual flow (see Section 2.6.5);

(iv) *haemolytic anaemia*, which occurs when erythrocytes rupture and release their cell contents into the plasma – so losing their oxygen-carrying capacity. The best example of haemolytic anaemia is sickle cell disease, an inherited blood disorder that produces erythrocytes with abnormal haemoglobin. The membrane of the erythrocytes becomes stiff and distorted in shape (forming a sickle or crescent); this makes it difficult for the cells to pass through the capillaries, with the result that some of them are ruptured in the process.

A severe form of anaemia is β-thalassaemia, an inherited disorder that is due to a defect in the synthesis of the β-globin chains of haemoglobin (see Section 3.2.4). This results in blood with very poor oxygen-transport capacity. Individuals with β-thalassaemia have an enlarged liver and spleen and require frequent blood transfusions – see below. (The word 'thalassaemia' is derived form the Greek *thalassa*, meaning 'sea', and -*aemia*, denoting 'blood'. It is so called because of the prevalence of the condition in eastern Mediterranean regions and the Middle East.)

Another important disorder affecting the blood is **leukaemia**. This disease is a cancer of the bone marrow with an uncontrolled proliferation of abnormal or immature white cells; the result is a greatly reduced ability to fight infectious diseases. Subsequent overwhelming infections are the most common cause of death in patients with leukaemia.

(b) Blood transfusions and blood groups

We now know that the surface membranes of human erythrocytes possess inherited **antigens** (unique surface recognition molecules) which define the specific blood group (or type) of an individual. Unfortunately, early attempts to restore a heavy loss of blood by the transfusion of blood from another person were frequently disastrous, with death a common outcome. The transfused cells would aggregate together into large clumps, a process called **agglutination**, which occurred when the blood types or groups of the individual donor and recipient were incompatible. The result was that small blood vessels in major organs of the recipient would became irreversibly blocked (reducing nutrient and oxygen supply to cells and tissues), with gradual organ failure being a possibility.

In the main blood group system, the **ABO system**, the erythrocytes of type A blood have A antigens; those with type B blood have B antigens; those with type AB blood have both A and B antigens; whilst those with type O blood do not have any A or B erythrocyte surface antigens. As mentioned above, if a person is given a blood transfusion of an incompatible type, adverse *antigen–antibody interactions* occur that can sometimes lead to fatal transfusion reactions.

Additional to the ABO system are 12 other erythrocyte antigen systems, the most important being **Rh factor**. People who have the Rh factor (so called because this antigen was first observed in rhesus monkeys) are defined as rhesus (Rh)-positive, whereas those without the Rh factor are Rh-negative. Most people (85% of the population) are Rh-positive, with 15% being Rh-negative. Unlike the ABO system, no naturally occurring antibodies develop against the Rh factor. However, Rh-negative individuals produce anti-Rh antibodies when they are first exposed to the alien Rh antigen in Rh-positive blood. Subsequent transfusion of Rh-positive blood to a sensitized Rh-negative person may produce a transfusion reaction. Rh-positive individuals, in contrast, never produce antibodies against the Rh factor that they themselves possess. Therefore, Rh-negative people are given only Rh-negative blood, whereas Rh-positive people can receive either Rh-negative or Rh-positive blood transfusions.

In practice, the compatibility between blood donor and recipient is determined simply by mixing the erythrocytes from the potential donor with plasma from the recipient. If no agglutination occurs then transfusion can be safely carried out.

2.3.2 The heart – a pump for life

The heart is the principal organ of the circulatory system, and is responsible for maintaining the flow of blood around the body. A human heart is about the size of a fist. The heart is a large four-chambered muscular bag, lying obliquely in the thoracic cavity, and is capable of increasing its pumping activity when a greater blood supply is required (Figure 2.1, overleaf). In order to understand how the heart works, we need to remind ourselves of the *primary* function of the cardiovascular system: the delivery of O_2, nutrients and other metabolic requirements to living cells throughout the body and the removal of CO_2 and other waste products from them.

When we breathe in, oxygen enters the main organ of the respiratory system – the lungs. Approximately 21% of the air we breathe is O_2 (see Chapter 3). In order to collect this inspired O_2 and release its waste cargo of CO_2, the blood has to be pumped around the lungs by the heart. **Oxygenated** (oxygen-rich) **blood** from the lungs returns to the heart and is then pumped around the body. Blood returning from the body tissues to the heart is rich in CO_2, produced by cellular respiration and depleted in oxygen. This blood is termed **deoxygenated** (oxygen-poor) **blood** and is pumped through the lungs to release CO_2 and collect more O_2. The design of the heart and associated blood vessels ensures that blood going to the lungs is kept separate from that going around the body

- Why do you think the separation of oxygenated from deoxygenated blood is important?

- It is essential not to mix deoxygenated and oxygenated blood, in order to maintain an adequate supply of O_2 to the tissues. If they were allowed to mix, then the blood O_2 content would be reduced.

How does the design of the heart maintain this separation between the flow of blood through the lungs and that around the body? The circulatory system actually consists of two separate circuits, the **pulmonary circulation** (heart → lungs → heart; i.e. to the lungs) and the **systemic circulation** (heart → body → heart; i.e. to the body) operating in parallel. These two separate circuits represent a 'double circulation system'. By virtue of its unique chambered design, the heart is able to serve both circuits at once, simultaneously pumping blood from one circuit through one half of its structure and blood from the other circuit through its other half (Figure 2.1) – as we shall learn, the harmonious operation of this double circulatory system is of great functional importance.

The musculature of the heart is called the **myocardium** (*myo* means 'muscle'; *cardium* means 'of the heart'); the structure and function of cardiac muscle are considered further in Section 2.4.1. Anatomically, the left and right sides of the heart are separated by a muscular wall, the *septum*, and each side is divided into a small

Figure 2.1 Gross anatomy of the heart: a cross-section through the heart showing the atria, ventricles and major vessels.

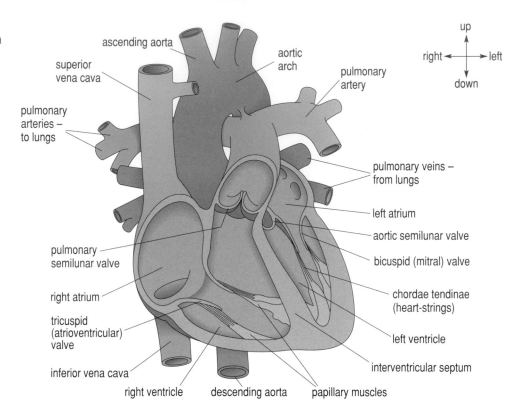

ascending aorta

aortic arch

superior vena cava

pulmonary artery

pulmonary arteries – to lungs

up

right ← → left

down

pulmonary veins – from lungs

left atrium

aortic semilunar valve

bicuspid (mitral) valve

pulmonary semilunar valve

right atrium

tricuspid (atrioventricular) valve

chordae tendinae (heart-strings)

left ventricle

interventricular septum

inferior vena cava

right ventricle   descending aorta   papillary muscles

chamber, the **atrium** (pl. *atria*), and a larger chamber, the **ventricle** (pl. *ventricles*) (Figure 2.1). The atria are connected to the ventricles through a circular channel with a valve that ensures a one-way flow. Deoxygenated blood returns from the body through two main 'great' veins, the **inferior and superior venae cavae** (Figure 2.2). These veins drain into the right atrium, a thin-walled chamber, which expands with little resistance as the blood enters. Blood from the right atrium flows into the right ventricle which is separated from the atrium by the *tricuspid valve* (also known as the *atrioventricular valve*, and shown in greater detail in Figure 2.3a). The three-cusped valve receives support from strong tendinous cords or *chordae tendinae* ('heart-strings') which pass from papillary muscles at the base of the ventricles to the ventricular surface of the valve flaps (Figure 2.1). The cords act like the guy ropes of a tent, preventing the valves from being inverted into the atria by the pressure within the ventricles. Thus the valve operates as a swing door which will only open in one direction. When blood enters the right atrium, the valve is open and blood flows into the right ventricle. When the ventricles contract the back-pressure of the blood forces the valves to close. Under normal conditions, contraction of the papillary muscles holds the cords in place to prevent any backflow into the atria.

● Why is it important that the blood is prevented from flowing back into the right atrium by the tricuspid valve?

● Without a tricuspid valve, contraction of the heart would force the blood back into the right atrium and into the main veins (venae cavae). The blood would be mixed with blood returning to the heart from the body, and this would prevent a one-way flow of blood out of the heart and into the lungs and reduce the pumping efficiency of the heart.

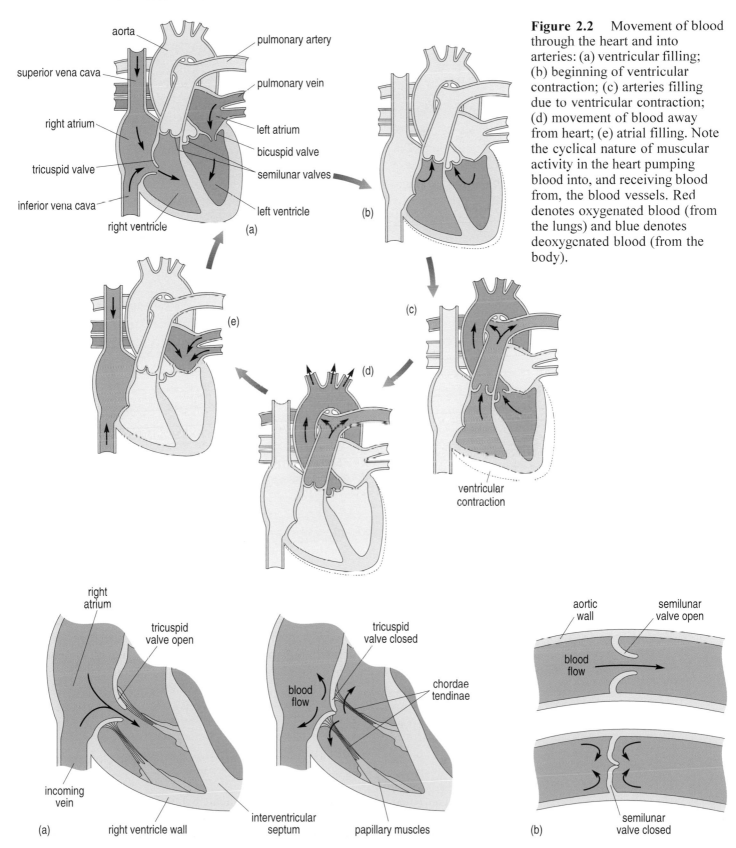

Figure 2.2 Movement of blood through the heart and into arteries: (a) ventricular filling; (b) beginning of ventricular contraction; (c) arteries filling due to ventricular contraction; (d) movement of blood away from heart; (e) atrial filling. Note the cyclical nature of muscular activity in the heart pumping blood into, and receiving blood from, the blood vessels. Red denotes oxygenated blood (from the lungs) and blue denotes deoxygenated blood (from the body).

Figure 2.3 Diagrammatic representation of valve mechanisms for (a) tricuspid and bicuspid valves and (b) semilunar valves.

All blood leaving the heart is pumped into arteries; the blood leaving the right ventricle is pumped into the **pulmonary artery**, which is the *only* artery in the body that carries deoxygenated blood. This artery branches into two smaller vessels, one serving each lung (*pulmonary* means relating to the lungs). Blood is prevented from draining back into the ventricles by a one-way valve in the pulmonary artery. The cusps of the valve are half-moon-shaped, so it is called a *semilunar valve* (Figure 2.3b). There is also a semilunar valve in the other main artery leaving the heart, the **aorta**. When the heart contracts, blood is forced out of the right ventricle into the pulmonary artery and the *pulmonary semilunar valve* is forced open. At the end of the contraction phase, the ventricle relaxes and the pressure in it temporarily falls below that in the pulmonary artery, causing a momentary reverse flow of blood. The valve closes and prevents blood draining back into the ventricle. Blood from the pulmonary artery passes around the blood vessels in the lungs, where it releases CO_2 and some water vapour and collects O_2. The water vapour keeps the membranes in the lungs moist – an important property of an efficient gas-exchange surface where the gases must be in solution (i.e. dissolved) when they enter the body. The oxygenated blood from the lungs passes back through the heart before it is pumped around the rest of the body. Blood from the lungs returns to the left side of the heart through the four large **pulmonary veins** (two from each lung) and enters the left atrium.

● What is the difference between the contents of the pulmonary veins and the contents of all other veins?

● The pulmonary veins carry oxygenated blood; all the other veins carry deoxygenated blood.

From the left atrium, blood flows into the left ventricle through the *bicuspid valve* (also called the *mitral valve*). When the heart contracts, the blood from the ventricle is forced into the main artery, the aorta, and leaves the heart. This blood is oxygenated and it is this blood that then flows around the rest of the body to the various organs and tissues via the systemic circulation. Again, any flow of blood back into the atrium is prevented by the bicuspid valve. Backflow of blood from the aorta to the ventricle is prevented by another semilunar valve, the *aortic semilunar valve*. Both the ventricles have much thicker walls than those of the atria, particularly the left ventricle. Contraction of the ventricular muscles is responsible for the main pumping action of the heart and the delivery of blood to the pulmonary and systemic circuits – the two parts of the double circulation system.

● What do you suppose is the advantage of a double circulation system?

● Oxygenated and deoxygenated blood is kept separate. Blood in the pulmonary vein is fully oxygenated, ready to provide other tissues with this essential molecule, but it will be at a lower pressure than arterial blood. The oxygenated blood is returned to the heart to be pumped out to the rest of the body.

The heart is encased in a double-walled sac filled with fluid, called the **pericardium** (*peri* means 'around'). The outer pericardial membrane is attached to the connective tissue that separates the two lungs and this attachment helps to keep the heart in its correct position within the chest cavity. The pericardial fluid provides lubrication between the walls of the pericardium and allows them to glide smoothly over each other during each heart beat. Inflammation of the pericardium is called *pericarditis*.

2.3.3 The vasculature – an interconnected system of pipes

It may be helpful at this point to revise details of the sympathetic and parasympathetic parts of the autonomic nervous system covered in Book 2, Chapter 1.

The vasculature forms a branching network of interconnected tubes which become progressively smaller and more branched with increasing distance from the heart. The first set of vessels, the arteries, carry oxygenated blood.

● Do all arteries carry oxygenated blood?

● No. The exception is the pulmonary artery, which carries deoxygenated blood from the heart to the lungs.

The major artery leaving the heart and delivering blood to the systemic circulation (i.e. to the body) is the aorta. The walls of the aorta and other arteries contain a large amount of elastic tissue (Figure 2.4a overleaf). When blood is pumped into the aorta, the elasticity of the artery walls allows the aorta to stretch or distend. When the heart relaxes, the stretched portion of the artery recoils, just like an elastic band, and forces the blood to move along the vessel with a pressure that is mainly defined by the physical parameters of the vessel through which the blood flows (see Section 2.7). In this manner, a wave of distension and constriction is created in the arteries, which is known as the **pulse wave**.

The walls of the aorta are approximately 2 mm thick and they form a tube which is about 2.5 cm in external diameter. The aorta feeds into the smaller arteries, which have thinner walls, approximately 1 mm thick, and an external diameter of about 0.4 cm. The arteries branch and divide into progressively smaller vessels. When the walls of the arteries are only some 30 microns thick (1 micron = 1 μm = one millionth of a metre – 0.000 001 m) and the diameter of the tube about 20 μm, the vessels are called arterioles. The walls of the arterioles consist mainly of smooth muscle; they can contract and relax in response to innervation from the nervous system or substances carried in the blood. Stimulation by **noradrenergic** nerve fibres (sympathetic nerves using noradrenalin as a neurotransmitter; see Book 2, Section 1.8.2) leads to contraction of the smooth muscle, whilst **cholinergic** nerve stimulation (parasympathetic nerves using acetylcholine as a neurotransmitter) causes the smooth muscles to relax. Whereas most blood vessel walls receive only sympathetic innervation, blood vessels in the gastrointestinal tract, salivary glands, endocrine glands and genital erectile tissues, as well as blood vessels in the brain, also receive parasympathetic innervation.

● What do you think is the consequence of contraction of the smooth muscle lining the arterioles?

● Contraction of the smooth muscle causes constriction of the arterioles.

Figure 2.4 (a) Blood vessel wall composition showing relative wall thicknesses and lumen sizes. Note the relative sizes are not to scale. Also shown are the relative compositions of the vessel walls. (b) High magnification image taken with an electron microscope showing three erythrocytes passing through the blood vessel wall of a small capillary. Their route will take them from the lumen of the capillary (top right of image), between two endothelial cells (aligned top left to bottom right) and into the tissue surrounding the capillary (bottom left of image). Note the presence of mitochondria in one of the endothelial cells and the biconcave shape of the erythrocytes. (The image has been coloured to highlight structural detail.) Magnification: 10 000 ×.

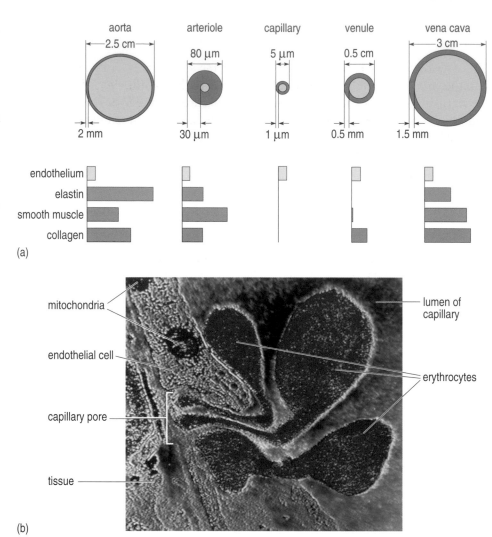

● Why is the capacity to constrict or dilate blood vessels important?

● To allow control of the flow of blood through specific blood vessels or vascular networks. Constriction of arterioles increases their resistance and reduces the amount of blood flowing through them, thereby causing an increase in overall blood pressure (see below). The reverse situation occurs if the arterioles are dilated, i.e. a decrease in arteriolar resistance and a decrease in blood pressure, thereby allowing the passage of a greater volume of blood.

Since arterioles are very narrow and their vascular network so extensive, and because they can be stimulated to constrict or dilate, they provide the major resistance to blood flow through the body. Resistance is a measure of the opposition to the flow of blood through a vessel, which is caused by friction between the moving fluid and the stationary vascular walls. Any small change in lumen diameter causes a proportionally large change in the resistance of the vessel. Fluid passes more readily through a large-diameter vessel than one with a smaller diameter. Imagine pouring water from a small kitchen jug into a common garden hose pipe compared with the much wider bore of the hose pipe used by firemen. The term **peripheral resistance** refers to the resistance of the systemic circulation to blood flow (see also Section 2.4.7). Consequently, blood vessels of the arteriolar system are sometimes referred to as the *arteriolar resistance vessels*.

The arterioles divide into smaller vessels with muscular walls, which in turn feed into the smallest vessels in the circulatory system, the **capillaries**. The diameter of the capillaries is only about 5 μm and their walls are formed by a single layer of endothelial cells; these are extremely thin wafer-like cells which have wavy edges and are laid end to end to form delicate lining membranes approximately 1 μm thick (Figure 2.4b). In fact, the capillaries are so small that even the blood cells have to squeeze through these vessels in 'single file'. Although the overall total area of the capillary walls in the body exceeds 6300 square metres (equivalent to the surface area of 11 tennis courts!) the capillaries contain, on average, only 5% of the total blood volume.

● The capillaries are the site at which exchange of materials and gases between the blood and the tissues occurs. What function is served by the 'thinness' of the capillary walls and their very small lumen diameters?

● The capillaries need to be narrow-bore vessels with thin walls to enable this exchange (most of which is by diffusion) to occur most efficiently. Also, because the capillaries are so abundant, they provide a large surface area over which exchange can occur.

The capillaries form vast networks of interconnected vessels, known as capillary beds, which supply all parts of the body with nutrients and remove the metabolic waste products.

Capillaries have an additional important function. As you will learn in Chapter 4, when tissue becomes damaged or invaded by foreign bodies, or both, the permeability of the capillary walls in the affected tissue is increased by a series of highly interrelated events – see Section 4.4.4. The spaces between the endothelial cells in the capillary walls, called the **capillary pores** (see Figures 2.4b and 2.11), are enlarged so that blood cells and plasma proteins can then pass from the capillaries into the affected tissue to begin repair or initiate defence against the invaders (see Figure 2.4b).

● Figure 2.4b shows erythrocytes passing between the endothelium of a capillary wall. Since white cells have a totally different function and are larger than erythrocytes, do you think they can also pass through the walls of capillaries?

● White cells are the front line soldiers in the fight against invading particles and alien bodies (see Chapter 4). Their presence at sites of tissue injury and potential infection are crucial, so although twice the size of erythrocytes, they too are able to pass into the tissue by squeezing through capillary pores.

Blood normally travels through the capillary beds and then into the extensive system of vessels, called the venous system, leading back to the heart. On entering the venous system, O_2 has passed from the blood into the tissues by diffusion (and CO_2 has diffused from the tissues into the blood), so the venous system carries deoxygenated blood loaded with CO_2. The venous system receives blood from the tissues and returns it to the heart. The **venules** are the smallest vessels of the venous system and collect blood from the capillary networks. Venules drain into the veins and eventually into the venae cavae and back to the heart. The walls of the venules and veins are much thinner and contain less smooth muscle and elastic tissue than the corresponding arterial walls. However, they have a larger internal diameter and are very distensible; in fact, during a complete blood transfusion the venous system can accommodate 90% of the transfused blood! Blood vessels of the venous system are therefore commonly referred to as *venous capacitance vessels* because of their great capacity to act as a peripheral blood reservoir.

The activity of skeletal muscles play a very important role in maintaining the flow of blood returning to the heart through the venous system (called **venous return**).

● How do skeletal muscles affect venous return?

● Many large veins, particularly those in the extremities, lie between skeletal muscles. When these muscles contract, the veins running through them are compressed. This inward compression forces blood to move along the veins in the direction of the heart, thereby aiding venous return (see Book 2, Section 4.7.1). This return pumping action is known as the 'skeletal muscle pump'. Blood flow through the venous system largely depends on the action of this special pumping mechanism.

The backflow of blood down veins due to gravity (especially in the lower limbs) is prevented by the presence of numerous one-way valves along the length of the veins (see Book 2, Figure 4.20).

Because there is some smooth muscle in the walls of veins and venules, they can also be stimulated, like arteries, to contract by the action of noradrenalin derived from sympathetic nerves. Contraction of the smooth muscle in the venous system causes constriction of the vessels, known as **vasoconstriction**. This reduces the movement of blood through the veins and reduces the venous volume. In contrast, relaxation of the smooth muscle causes dilation of the vessels, called **vasodilation**, and this increases the blood volume. (Note here, that the terms *vasoconstriction* and *vasodilation* can be applied to any blood vessel.) Of specific functional importance is that the venous system can adjust the total capacity of the circulation to meet the requirements of the body and regulate the blood volume returning to the heart, thereby influencing the amount of blood the heart has to pump and therefore the work the myocardium has to perform.

Summary of Section 2.3

1 Blood is composed of erythrocytes, leukocytes, and platelets suspended in plasma – all with their own function. Erythrocytes have surface recognition molecules, called antigens, which define a person's blood group. The percentage of a given volume of blood occupied by erythrocytes is called the haematocrit ratio and is of diagnostic value for evaluating blood disorders such as anaemia. Leukaemia is a cancer that affects the bone marrow and the production of leukocytes.

2 The heart is a four-chambered, muscular pumping bag. The right-hand side of the heart receives deoxygenated blood from the body through the superior and inferior venae cavae, and pumps it via the pulmonary artery to the lungs where it becomes oxygenated. The blood is returned to the left side of the

heart through the pulmonary veins (pulmonary circulation). The oxygenated blood is pumped out of the heart via the aorta to the rest of the body (systemic circulation).

3   Blood is transported around the body in an interconnected network of blood vessels called the vasculature. Blood flows from arteries, to arterioles, to capillaries (the smallest vessels of the circulation), to venules and into veins.

4   Blood is pumped through the vasculature as an advancing pulse wave with a given pressure. Changes in the diameter of the blood vessels greatly affect the peripheral resistance of the vasculature to blood flow. This can be altered by the contraction or relaxation of smooth muscles in arterial walls. Veins can dilate to store a substantial volume of blood, thereby influencing the volume of blood returning to the heart.

5   The flow of blood through the heart and through the venous system is controlled by one-way valves and by innervation from the autonomic nervous system.

2.4 Normal function of the cardiovascular system

We discuss next how the heart and the cardiovascular system function normally. This will provide the basis for subsequently understanding what happens when things go wrong.

2.4.1 How does the heart work? The beat goes on, and on...

As mentioned above, the heart is basically a four-chambered muscular pumping bag. But how does it beat and pump blood? You may have seen in horror films a heart that has been freshly ripped out of a body beating on its own. This is not pure fantasy, for a heart will continue to beat for several hours if it is removed carefully from the body and kept in a solution containing nutrients and O_2. This is possible because the muscular activity of the heart, which generates the heartbeats, begins within the heart muscle itself and can continue without a nerve supply. The heart has its own innate or intrinsic, rhythm producing physiological mechanism that perpetuates its own activity.

So what generates this activity? The muscle fibres that make up the heart muscle (**cardiac muscle**) are similar, though not identical, to skeletal muscle fibres (Book 2, Section 4.7.1). For one thing they must never tire; even fatigue-resistant skeletal muscles get nowhere near their performance! Cardiac muscle cells are connected to each other by specialized membrane junctions (called gap junctions – see Figure 2.19 in Book 1) that permit the spread of electrical activity from one cell to the next. This ensures the rapid and uniform spread of contraction throughout the wall of the heart.

What initiates and controls the contraction? There is a small mass of highly specialized cardiac muscle cells embedded in the wall of the right atrium close to where the superior vena cava enters. This is called the **sinoatrial node (SAN)** and experiments have shown that this small mass of muscle cells function as the heart's **pacemaker**, which is responsible for its intrinsic rhythmic activity (Figure 2.5). If the SAN is cut out (excised) from the surrounding tissue, it will contract at about 80–85 beats per minute. Other pieces of excised atrial tissue will also beat on their own but at a slightly slower rate, and tissue from the walls of the ventricles will contract even more slowly, at about one-third of the intrinsic rate. Although the different parts of the heart can contract at their own innate rhythm, the contraction of the heart follows the activity of the SAN, called the **sinus rhythm** – this sets the normal rate of contraction for the whole organ.

Contraction of the heart is thus initiated by electrical activity in the SAN. This wave of electrical excitation is transmitted through the right and then the left atrium, leading to contraction of the atrial muscle fibres. When the wave reaches the junction between the atria and ventricles, it excites another group of specialized muscle fibres at the **atrioventricular node (AVN)** (Figure 2.5). Here the speed of electrical conduction slows, so transmission is delayed briefly, thus allowing the atria to complete their contraction before the ventricles contract. (This delay is called 'AVN delay' and is very important in coordinating the activity between the atria and ventricles.) The AVN is continuous with a bundle of numerous modified cardiac muscle fibres known as *Purkinje fibres*. The larger Purkinje fibres form a long bundle, called the *atrioventricular bundle* (also known as the *bundle of His*) which runs down the septum between the ventricles (interventricular septum; see Figure 2.5). The speed of electrical impulse conduction increases along the bundle of His so that the impulse quickly reaches the Purkinje fibres that divide and fan out in a network over both ventricles at the base of the heart (Figure 2.5). When an electrical impulse reaches the ends of the Purkinje fibres, it spreads through the ordinary cardiac muscle fibres and this leads to contraction of the ventricles. The impulse first stimulates the cardiac muscle at the base of the heart, which contracts from the bottom of the ventricles upwards, forcing the blood upwards and out of the heart. Thus, the pacemaker of the heart, the SAN, sends out rhythmical waves of electrical excitation which are transmitted over the atria, through the AVN and Purkinje fibres to the ventricles (Figure 2.5).

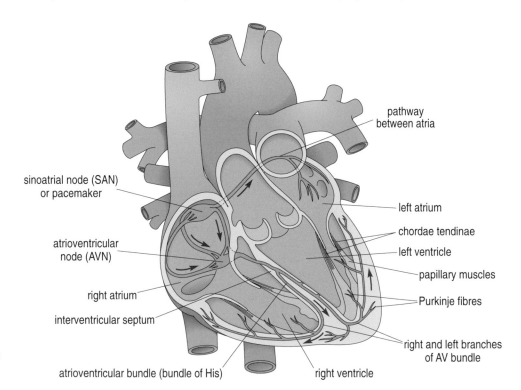

Figure 2.5 Transmission of electrical activity (red arrows) through the heart, which triggers contraction.

One complete heartbeat takes about 0.85 seconds and is known as a **cardiac cycle**. It consists of a period of contraction, **systole**, and a period of relaxation, **diastole**. You met the terms systolic pressure and diastolic pressure in Book 1 (Section 1.5.1). Indeed, blood pressure is defined as the ratio of these two pressure readings; for example, a blood pressure of 120/75 means that the systolic pressure during contraction is 120 mmHg and the diastolic pressure during relaxation is 75 mmHg (see Section 2.4.8 for more details).

2.4.2 The electrocardiogram (ECG)

The condition of the heart can be monitored by studying the ECG, which gives information about the transmission of the electrical activity through the heart.

The muscular contraction of the heart is preceded by a wave of electrical currents similar to a nerve impulse – called a cardiac action potential (see Book 2, Section 1.9.1). Since body fluids are good conductors of electricity, this electrical pattern or wave of depolarization (a reduction in the size of the electric potential difference across the cell membrane) spreads through the tissues and can be detected using electrodes placed on the skin at specific positions around the heart. The electrical activity can be amplified, recorded and an ECG subsequently charted to give a record of the spread of electrical activity through the heart. In practice, ECG recordings are routinely made with 10 electrical leads. These are placed in a standard arrangement: six chest leads at defined sites across the heart and the remaining leads at defined positions on the limbs. Such an arrangement of recording electrodes makes it possible to detect how and when electrical activity is transmitted through individual parts of the heart. The 10 electrical leads produce 12 'views' of the electrical activity across the heart. This information is of great clinical importance since it allows an immediate diagnosis of abnormal heart muscle function – for example, abnormal electrical activity, irregular heart rates or damage to the heart muscle (see Section 2.6).

A typical trace of a normal adult ECG recorded through lead II is shown in Figure 2.6a and b (overleaf). (Lead II is chosen because it shows the heart's electrical activity along the interventricular septum – the major axis of the heart; see Figure 2.5). Examine these diagrams carefully and identify the various components of the normal ECG wave; consider the shape of the traces, the order and frequency of various components. The normal ECG is composed of a P wave, a 'QRS complex' and a T wave. In some people, there is also a smaller U wave, but this is not always seen.

Each section of the ECG trace corresponds to a particular phase during the cardiac cycle, i.e. the sequence of physical events taking place in one cycle of the heart's action. Importantly, the electrical activity associated with each phase can be 'viewed electrically' from the various electrode positions across the heart. The P wave is produced by depolarization of the atria, prior to atrial contraction. This is followed by the 'QRS complex' which indicates both atrial repolarization and ventricular depolarization prior to ventricular contraction, and finally the T wave, which indicates repolarization of the ventricles, i.e. the ventricular muscle cells returning to a relaxed state. In those cases where the U wave can be observed, it is suggested that this corresponds to the slow repolarization of the papillary muscles which attach the chordae tendinae to the ventricle wall (Figure 2.1).

Figure 2.6 Basic components of a typical normal adult electrocardiogram (ECG) as recorded through electrical lead II (see text) are defined in (a) and displayed in the trace (b). Traces (c), (d) and (e) show abnormal ECG recordings. (c) Atrial fibrillation; individual P waves are replaced by multiple P waves (black arrows) and there is an irregular ventricular QRS complex rhythm. (d) Ventricular fibrillation, which is characterized by a highly irregular waveform, no clearly identifiable QRS complexes or P waves, and an unstable baseline. (e) Complete heart block, where the atria and ventricles are electrically disconnected: although P waves (black arrows) occur regularly, they are not conducted to the ventricles and show no obvious relation to the QRS complexes.

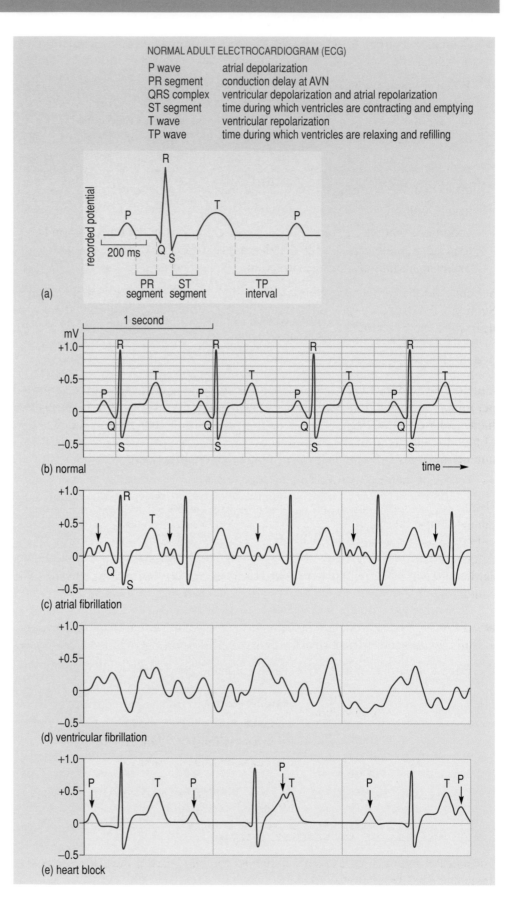

NORMAL ADULT ELECTROCARDIOGRAM (ECG)

P wave	atrial depolarization
PR segment	conduction delay at AVN
QRS complex	ventricular depolarization and atrial repolarization
ST segment	time during which ventricles are contracting and emptying
T wave	ventricular repolarization
TP wave	time during which ventricles are relaxing and refilling

(b) normal

(c) atrial fibrillation

(d) ventricular fibrillation

(e) heart block

● Critically examine the abnormal ECG traces in Figures 2.6c–e. What might be possible underlying causes for the different waveforms?

● When compared with the normal ECG trace (b), the abnormal ECG traces are irregular with their own characteristic waveforms. Such waveforms result from uncoordinated patterns of electrical activity in the heart (termed **cardiac arrhythmias**; see Section 2.6.4) and may result from one of a number of causes:

• poorly synchronized or unsynchronized electrical activity;

• drugs that affect propagation of the cardiac action potential;

• heart disease;

• conditions that alter K^+ and Ca^{2+} ion concentrations in the body (K^+ and Ca^{2+} ions are crucially involved in the electrical activity of the heart and the mechanisms underlying its contraction).

We consider the cardiac arrhythmias illustrated in Figures 2.6c–e in more detail in Section 2.6.4. Next we turn our attention to the amount of blood pumped out of the heart during each cardiac cycle.

2.4.3 Cardiac output

Cardiac output is a measure of the volume of blood pumped out of the ventricles per minute. In effect, it is a measure of the total blood flow per minute through the lungs and around the body. The calculation of cardiac output depends on the volume of blood pumped out of the heart during one beat, the **stroke volume**, and the number of times the heart beats per minute, i.e. the heart rate.

cardiac output = stroke volume × heart rate

Heart rate is determined simply by measuring the pulse, but the volume of blood pumped out of the heart with each beat is not measured so easily. Thus, the cardiac output is measured directly and, by simple calculation, the stroke volume is determined. There are established experimental techniques available to measure cardiac output; however, the details are complex and are beyond the scope of this course.

● If the average heart rate is 70 beats per minute and cardiac output is 5 litres per min, what is the average stroke volume in millilitres (ml) of blood?

● Equation: cardiac output = stroke volume × heart rate

$$\text{stroke volume} = \frac{\text{cardiac output}}{\text{heart rate}}$$
$$= \frac{(5 \text{ litres per min})}{(70 \text{ beats per min})} = \frac{5 \text{ litres}}{70 \text{ beats}}$$
$$= \frac{5000 \text{ ml}}{70 \text{ beats}} = 71 \text{ ml of blood per beat (approx.)}$$

2.4.4 Alterations in cardiac output

The cardiac output is not a fixed volume but can be altered if the body needs a greater or a reduced supply of blood.

● Can you think of any situations where there may be a need for an increased supply of circulating blood?

● You may have thought of a number of different situations, but the most common is during exercise. Exercising muscles need a greater supply of blood than do resting muscles.

The effects of a variety of different conditions on cardiac output are summarized in Table 2.2.

Table 2.2   Effect of various conditions on cardiac output. Approximate percentage changes from normal resting condition (lying flat) are shown in brackets.

Condition or factor	Effect
sleep	no change
moderate changes in environmental temperature	no change
anxiety and excitement	increase (50–100%)
eating and digestion	increase (30%)
exercise	increase (up to 70%)
high environmental temperature	increase
pregnancy (late)	increase
sitting up or standing up (the initial effect)	decrease (20–30%)
rapid arrhythmia (heart beating irregularly)	decrease
heart disease	decrease

What brings about a change in cardiac output? From the equation defining cardiac output in Section 2.4.3, it can be seen that a change in the stroke volume and/or heart rate will both alter cardiac output.

2.4.5 Control of stroke volume

Stroke volume, the amount of blood pumped out with each contraction, depends on the amount of blood that returns to the heart from the veins, i.e. the venous return. If a large volume of blood returns to the heart, a large volume of blood will be pumped out of the heart. This is because the strength of contraction (or *inotropy*) of the heart depends on the initial length of the muscle fibres in the atria. If the venous return is large, a large volume of blood will enter the right atrium during diastole when the heart is relaxed. The muscle fibres in the atrium stretch to accommodate the greater volume of blood, and the systolic contraction is greater, thereby increasing the stroke volume. This is **Starling's law**, discovered by the eminent physiologist Ernest H. Starling (1866–1927). Since the two sides of the heart (forming the double circulatory system, mentioned previously) work in parallel, the same amount of blood will be pumped out of each ventricle.

● Can you think of any situations where the blood volume returning to the heart would be increased?

● During exercise, the skeletal muscles are contracting and squeezing the blood vessels running through them. This increases the blood pressure and the volume of blood returning to the heart.

Athletes who train hard can increase the length of their heart muscle fibres by as much as 50%. Since the atrial muscle fibres can stretch further than those of untrained people, the venous return is consequently increased. An increase in the volume of venous blood returning to the heart produces an increase in stroke volume, consequently the cardiac output of a trained athlete is also greater than that of an untrained person.

2.4.6 Control of heart rate

Although the heart rate is intrinsically set by the heart's internal pacemaker, the SAN, it is also carefully regulated by the nervous system (see Book 2, Section 1.8.2 and especially note Figure 1.16). The SAN is innervated by both sympathetic and parasympathetic nerves which originate in the two cardiac centres of the brain that are responsible for regulating heart rate. The sympathetic nerves originate in the **vasomotor centre**, located in the medulla of the brain, but reach the heart via the spinal cord as shown in Figure 2.7 (overleaf). Sympathetic nerves use the neurotransmitter *noradrenalin* to stimulate (excite) the heart. In contrast, the parasympathetic nerves originate in the **cardio-inhibitory centre**, also situated in the medulla, and descend to the heart via the 10th cranial nerve called the *vagus nerve* (see Figure 1.11 in Book 2). The SAN is under **tonic** (sustained) **inhibition** from the parasympathetic vagus nerve, which uses the neurotransmitter acetylcholine to affect the heart and other structures. This means that the activity of the SAN is continuously held 'in check' by cholinergic vagal input (called *vagal restraint*). In other words, if the vagus nerve to the heart was cut, the heart rate would increase due to the lack of vagal restraint.

● In a normal person what do you think happens if there is an increase in the activity of the vagus nerve?

● Vagal restraint is increased and heart rate is slowed even further.

A slowing of the heart rate, by whatever means, is called **bradycardia**. For example, shock or fear can cause reflex bradycardia as a result of increased vagal restraint activity – the classic 'heart-stopping' scenario after a shock! Conversely, an increase in heart rate is called **tachycardia** – this can be produced by an increase in the activity of the sympathetic noradrenergic (noradrenalin-releasing) nerves from the brain, which directly increase the activity of the heart's SAN.

Heart rate is also altered in response to changes in blood pressure (see Section 2.4.7). Receptors that detect alterations in blood pressure are called baroreceptors (Section 1.5). Baroreceptors are found in a small protrusion from the carotid artery (one of the main arteries, which delivers blood to the brain; see Figure 2.9 later) called the *carotid sinus*, and in the walls of the aorta (see Figure 2.12 later). These receptors detect small changes in blood pressure in the carotid artery and aorta. If a larger volume of blood has been pumped out of the heart due to an increase in the venous return, the blood pressure in these arteries will increase. The baroreceptors detect this increase and send the information to the cardio-inhibitory and vasomotor centres in the medulla of the brain (Figure 2.7). In the cardiac centres, the information is processed by interneurons and returned to the SAN via the sympathetic and parasympathetic nerves. If the blood pressure increases, the cardiac centres stimulate activity in the vagus nerve by a reflex action to produce bradycardia; this results in a reduced cardiac output and thus a decrease in blood pressure.

● What do you think happens if the baroreceptors detect a decrease in blood pressure; for example, when someone stands up quickly (see Case Report 2.3)?

● The baroreceptors report a decrease in blood pressure to the cardiovascular centres in the brain which, through local nerve circuits in the medulla, cause an increase in sympathetic stimulation of the SAN. The effect is to increase heart rate and cardiac output and thereby raise blood pressure back to a normal level.

Circulating hormones, such as adrenalin or noradrenalin, can also increase heart rate by their direct stimulatory activity on the SAN. This is the reason our heart rate increases during periods of stress since these are the so-called 'fright, fight or flight' hormones.

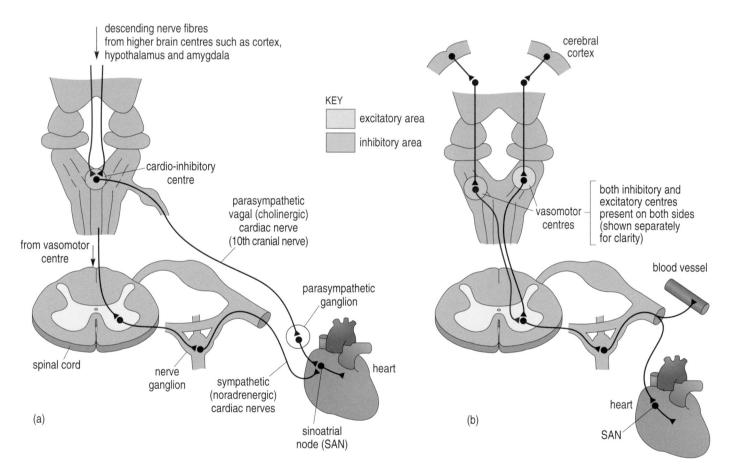

Figure 2.7 Schematic diagram showing innervation of the sinoatrial node (SAN) by nerves from the cardio-inhibitory centre (a) and vasomotor centres (b) in the brain. Inhibitory nerves/pathways are shown in red. In (a), descending nerve fibres from higher brain centres innervate nerve cells in the cardio-inhibitory centre that send parasympathetic nerve fibres in the vagus nerve, which directly innervate the SAN. In (b), descending nerve fibres contact neurons in the inhibitory and excitatory medullary vasomotor nerve centres. Neurons in these centres send axons to the spinal cord, which innervate a pathway of sympathetic nerve fibres that contact the SAN and blood vessels. Note that the SAN is innervated by both sympathetic and parasympathetic fibre systems, as shown in (a).

2.4.7 Pulse and blood pressure

Do you have a pulse? The subclavian arteries branch from the aortic arch (the region of the aorta that bends away from the heart) and continue as the axillary arteries inside the armpits (axillae) (see Section 2.5). The axillary arteries lead into the basilar arteries of the upper arm these in turn give rise to the ulnar and radial arteries in the forearm and wrist. On the palm side of your hands, the radial arteries travel close to the base of each thumb. Gently press two fingers over the radial artery in your wrist (at a point located about 1–2 cm below the bony base of your thumb) and count the number of pulses per minute. Having found your pulse, what's your pulse rate? At rest, the average rate is 50–80 beats per minute, with considerable variation between individuals. How does your pulse vary with different types of exercise? Or at different times of the day? Or when standing up or lying down?

The pulse you feel is a measure of the pulse wave, and relates to the alternate expansion and recoil of artery walls; in effect, your pulse rate is your heart rate. Systolic contraction forces blood into the aorta, the walls expand to accommodate the blood, and during diastole the elastic recoil of the walls moves the blood down the aorta in a wave (Figure 2.8). Every time the heart contracts, a pulse wave begins; it is these pulsations that you can feel in your arteries and which are counted to indicate your heart rate. The pulse wave is like a shock wave transmitted through the arteries ahead of the blood. The velocity of the pulse wave is about 4–5 metres per second whereas the velocity of blood flow is less than 0.5 metres per second.

Blood pressure is the force that is exerted by the blood on the surface of the inner walls of the blood vessels (see Section 2.3.3). The mean blood pressure is determined by the rate of blood flow and the resistance to blood flow. Blood flow in turn is directly dependent on the pumping activity of the heart and therefore on cardiac output. When cardiac output increases, blood flow increases and blood pressure increases. When cardiac output decreases, blood flow decreases and blood pressure decreases.

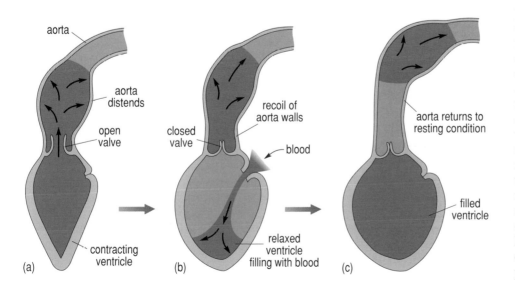

Figure 2.8 Development of the pulse wave. (a) Systolic contraction forces blood into the first section of the aorta which distends under the pressure. (b) The aortic valve closes as the aortic walls recoil and blood is forced further up the aorta, distending the walls. The ventricle again fills with blood. (c) The first section of the aorta returns to its 'resting' state and the next section is beginning to recoil and force blood further along the vessel with expansion of the vessel walls. The ventricle has filled again and is about to contract and initiate the next pulse wave.

● With your knowledge of the factors that affect cardiac output, can you suggest what activities or conditions might increase blood pressure?

● Exercise, which can increase cardiac output by up to 70%, will increase blood pressure, as will anxiety (see Table 2.2). Reduced elasticity of the artery walls affecting peripheral resistance will also increase blood pressure. This might occur due to a gradual 'hardening' of the arteries – a process called atherosclerosis (see Book 1, Section 3.4.1 and also Section 2.6.2 of this book).

Blood pressure also depends on the volume of blood flowing through the system. If the blood volume falls, e.g. due to extensive bleeding, the blood pressure will fall until homeostatic mechanisms have been activated to restore the normal pressure. If the blood volume increases, the blood pressure will increase. Some people who have a high intake of salt (dietary sodium) can experience problems with high blood pressure since excess Na^+ in the diet causes retention of water in the blood, which in turn increases blood volume and thus blood pressure (see Section 1.5.2 and Section 2.6.1).

The flow of blood, like that of any other liquid, is affected by the resistance of the system through which it flows. If the resistance increases, then the blood pressure increases. The factors that contribute to the resistance to blood flow through the systemic circulation, or *peripheral resistance*, are:

• the thickness (viscosity) of the blood;

• the length of blood vessels through which the blood flows;

• friction between the blood and the vessel walls.

In a healthy individual and under normal circumstances, the viscosity of the blood remains fairly constant and has little bearing on blood pressure. By far the most important factor in determining blood pressure is the friction between the blood and the vessel walls. Friction depends on the surface area of the vessel, which relates to the vessel's length and diameter. The length of blood vessels does not change, but vessel diameter decreases along the route from the aorta to the arteries and on to the arterioles.

● What factor do you think will cause the greatest change in the resistance of peripheral arterial vessels?

● A change in the diameter of the lumen of the arterioles.

Thus, if the walls of the arterioles contract, the size of the lumen will decrease, increasing the friction and greatly increasing peripheral resistance. There is a simple equation that allows the calculation of mean arterial blood pressure:

mean arterial blood pressure = cardiac output × total peripheral resistance

2.4.8 Measurement of blood pressure

Blood pressure in the brachial artery is normally measured by fitting an inflatable cuff closely around the upper arm at the level of the heart. The more traditional method involves attaching a portable glass column containing mercury (a mercury U-tube manometer) to the cuff; this device is known as a **sphygmomanometer**. The manometer measures the pressure in the cuff and this pressure can be altered by pumping air into the cuff and releasing it through a valve. The manometer is now being widely replaced by a small calibrated electronic pressure monitor. After inflating the cuff, a stethoscope is placed over the brachial artery inside the elbow joint. By releasing the pressure inside the cuff and listening to the characteristics sounds made by blood flowing through the partially compressed artery (called 'Korotkoff sounds') it is possible to determine the maximum and minimum blood pressures. The maximum pressure, which is reached during the period of ventricular contraction, is the systolic blood pressure, whilst the minimum pressure, which registers during ventricular relaxation, is the diastolic blood pressure (see Section 2.4.1).

Blood pressure is measured in units of mmHg (see Box 2.1). The systolic blood pressure is normally between 100 and 160 mmHg and the diastolic blood pressure is normally between 60 and 90 mmHg.

The difference between the systolic and diastolic blood pressure is called the *pulse pressure*. Blood pressure measurements vary greatly between individuals and reflect, not only the overall ability of the cardiovascular system to pump blood through the vast vascular network which permeates the entire body, but also the functional and structural characteristics of the nervous, renal and endocrine systems and their complex regulated interactions (see Section 2.4.1, also Chapter 1 in this book). Because so many intermediate systems contribute to a person's blood pressure, normal genetic variation in these systems is a possible source underlying the significant blood pressure differences between individuals as well as the subsequent risk of them developing abnormal cardiovascular function (see Section 2.6).

Box 2.1 Units of pressure

In physiology, the unit of pressure is conventionally expressed as millimetres (mm) of mercury (Hg). 'Millimetres of mercury' (mmHg) refers to the height of a column of mercury attached to an instrument that detects pressure (e.g. a U-tube manometer). The height of the mercury column inside the closed side of the U-tube is directly related to the pressure applied to the open end of the manometer.

There are other units of pressure in daily usage – for example, 'pounds per square inch' (lb sq in^{-1}, psi) used to measure car tyre pressures and 'millibars' (mbar) used in weather forecasts. All the units of pressure can be interconverted so, for example, the Earth's atmospheric pressure is 760 mmHg or about 15 lb sq in^{-1} or 0.013 mbar.

2.4.9 The control of blood flow

The control of blood flow is achieved by alteration of the peripheral resistance. As mentioned earlier, the diameter of the blood vessels is influenced by the activity of sympathetic (and, in some cases, also parasympathetic) nerves innervating the smooth muscle in the walls of the vessels. Vasoconstriction is stimulated by noradrenalin (sympathetic innervation), because this neurotransmitter reduces blood flow and increases peripheral resistance.

Vasodilation is stimulated by acetylcholine (parasympathetic innervation), which increases blood flow and reduces peripheral resistance. As you will learn in Section 2.6.1, vasodilation is a very important mechanism underlying the treatment of high blood pressure.

Blood vessels are also affected by a number of other circulating substances. These include adrenalin and noradrenalin (produced by the adrenal glands) and angiotensin II and antidiuretic hormone (ADH) which both directly cause vasoconstriction (see Section 1.5). Control of blood flow by circulating factors is known as **humoral control**. *Bradykinin* and *histamine* are such factors and are released by damaged cells (see Book 1, Section 2.10.2) as well as being activated during an allergic response (Chapter 4); they cause vasodilation which produces a local increase in blood flow through injured/damaged body areas.

The blood flow to all organs increases when they are metabolically active, thereby maintaining a continuous supply of O_2 and nutrients. The decreased O_2 concentrations and increased CO_2 concentrations that result from the initial raised metabolic rate, act as local vasodilators and increase the blood flow in proportion to the increase in metabolism; this is known as **metabolic autoregulation**.

Summary of Section 2.4

1   The contraction of the heart is initiated by electrical activity in the SAN, the heart's pacemaker. This activity is distributed around the heart by modified muscle fibres, producing atrial contraction followed by ventricular contraction. The period of contraction (systole) when blood is pumped out of the heart is followed by a period of relaxation (diastole) when the heart again fills with blood.

2   The condition of the heart can be monitored by listening to the heart sounds, which give information about the condition of the valves of the heart, or by studying the electrocardiogram (ECG), which gives important information about the normal and abnormal transmission of electrical activity through the heart.

3   Cardiac output is a measure of the volume of blood pumped out of the heart in one minute and depends on the stroke volume and the heart rate. Cardiac output can be increased when required, by an increase in either the heart rate or the stroke volume, or both.

4   Stroke volume is increased by an increase in venous return.

5   The heart rate is increased by a change in the activity at the SAN. This can result from a reduction in tonic inhibition, i.e. a reduced rate of firing of the vagus (parasympathetic) nerve, an increase in sympathetic stimulation, or an increase in levels of the circulating hormones (adrenalin and noradrenalin).

6   Changes in blood pressure are monitored by baroreceptors situated in the walls of the carotid artery and the aorta; information is sent from these sensors to the cardiac centres in the medulla, which cause either an increase or a decrease in the rate of contraction of the heart as appropriate.

7   Blood pressure is the pressure exerted by the blood on the walls of the vessels and is measured using a sphygmomanometer. It is expressed as 'systolic pressure' over 'diastolic pressure'.

8   Humoral control and metabolic autoregulation influence the diameter of arterioles, thus modifying peripheral resistance.

2.5 The systemic circulation

The flow of blood around the body (excluding the lungs) occurs through the systemic circulation. This is achieved mainly by the pumping activity of the heart, but it is helped by:

*   contraction of skeletal muscles, particularly during exercise (see Section 2.3.3);

*   the elasticity of the arteries;

*   the fall in pressure in the chest cavity that occurs during inspiration, which reduces the resistance of the blood vessels.

Some of the main branches of the systemic circulation are shown in Figure 2.9, and in the next few sections we will briefly discuss the circulation to and from selected organs and tissues.

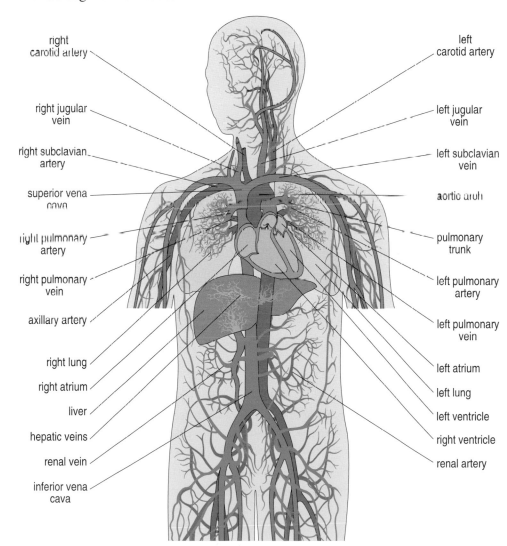

right carotid artery

right jugular vein

right subclavian artery

superior vena cava

right pulmonary artery

right pulmonary vein

axillary artery

right lung

right atrium

liver

hepatic veins

renal vein

inferior vena cava

left carotid artery

left jugular vein

left subclavian vein

aortic arch

pulmonary trunk

left pulmonary artery

left pulmonary vein

left atrium

left lung

left ventricle

right ventricle

renal artery

Figure 2.9   A summary diagram of the systemic circulation showing the location of the heart and major vessels, and their relationship with some of the major body organs. All arteries except the pulmonary artery (i.e. vessels carrying oxygenated blood) are shown in red; all veins except the pulmonary vein (i.e. vessels carrying deoxygenated blood) are shown in blue.

2.5.1 Coronary circulation – the heart's own blood supply

Although all the body's blood passes through the heart, the walls of the heart are so thick that the heart muscle requires its own circulation. Cardiac muscle is supplied by two coronary arteries, which branch off from the aorta as it leaves the heart. This supply accounts for about 4–5% of the output of the left ventricle. The coronary arteries branch within the walls of the heart, eventually forming a capillary bed through which O_2 and nutrients diffuse into the cardiac muscle cells. The coronary capillaries drain into the coronary veins and ultimately into the **coronary sinus**, which is a small valve-regulated opening that allows the blood in the coronary circulation to drain directly back into the right atrium.

● What would be the effect of a blockage in a coronary artery?

◐ A blockage in either one of the coronary arteries would deprive one area of the heart muscle of O_2 and nutrients. If the blockage completely impairs delivery of blood, then that area of the heart muscle could die through lack of oxygen.

Blockage or narrowing of the coronary arteries, as with obstruction of the arteries supplying all the major organs, are potentially life-threatening events and are considered below.

2.5.2 Cerebral circulation and the blood–brain barrier

The brain receives approximately 15% of the total cardiac output and due to its very high metabolic activity it also has a very high oxygen consumption – about 20% of the body total. This is exceeded only by the liver and is equivalent to that of actively contracting skeletal muscle. A continuous supply of O_2 and nutrients is essential since brain cells are particularly susceptible to damage if shortages of these occur. In situations of high demand in other regions of the body, there is no reduction in cerebral circulation to compensate – the brain acts as a tyrannical dictator, safe-guarding the available oxygenated blood for its own use.

Blood is delivered to each side of the brain by two sets of major arteries comprising the left and right internal carotid arteries and the left and right vertebral arteries. The ascending carotid arteries on both sides of the neck (Figure 2.9) branch to form the left and right, internal and external carotid arteries – only the left and right internal carotid arteries enter the cranial cavity (Figure 2.10). Once inside the cranial cavity, the internal carotid arteries give rise to side branches, called the ophthalmic arteries, which supply the eyes. In contrast, the left and right vertebral arteries arise from the left and right subclavian arteries, respectively (Figure 2.9), and ascend to the brain either side of the vertebral column (Figure 2.10). On approaching the brainstem, the vertebral arteries form the basilar artery (Figure 2.10).

At the base of the brain, the left and right internal carotid and basilar arteries are joined together by communicating vessels to form a vascular network (Figure 2.10), called the 'circle of Willis' – named after the famous physician Thomas Willis (1621–1675). At the 'circle of Willis' the internal carotid arteries give rise to the anterior and middle cerebral arteries which supply the frontal and middle parts of the cerebral hemispheres respectively. The basilar arteries lead into the posterior cerebral arteries (Figure 2.10) supplying the cerebellum, the brainstem and the posterior parts of the cerebral hemispheres. Of strategic functional importance is that all the blood entering the brain must first pass through the 'circle of Willis'.

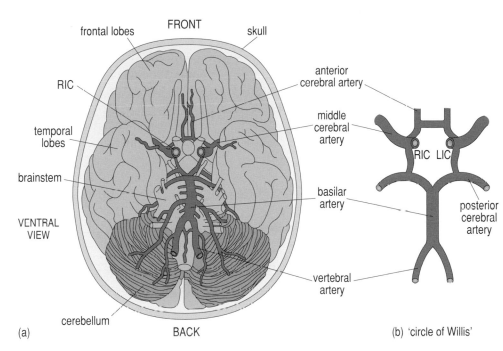

FRONT

frontal lobes   skull

RIC

temporal lobes

brainstem

VENTRAL VIEW

cerebellum

(a)   BACK

anterior cerebral artery

middle cerebral artery

RIC LIC

basilar artery

posterior cerebral artery

vertebral artery

(b) 'circle of Willis'

Figure 2.10 The 'circle of Willis'. (a) The position of the arterial circle underneath the brain. This is a view from beneath the brainstem showing the cut ends of the four arteries entering the brain. (b) The isolated arterial circle (simplified). RIC = right internal carotid artery; LIC = left internal carotid artery.

● What advantage does the 'circle of Willis' give cerebral blood supply?

● If one of the arteries serving the brain becomes blocked, flow of blood around the 'circle of Willis' could ensure that this blockage is compensated for by blood from another artery. In this way, the 'circle of Willis' acts as an 'arterial shunt' and provides a system of blood vessels that can deliver blood to all the brain cells.

Blood from the brain drains into the jugular veins and returns to the heart via the superior vena cava. The control of cerebral blood flow is the most highly developed and sophisticated in the body. Neural control is mainly via noradrenergic neurons in the cerebral cortex, which lead to vasoconstriction. The cholinergic input, which has a vasodilatory action, is derived from the cranial nerves that innervate the face. Much of the local vasodilation is due to metabolic autoregulation as a result of the O_2 demand of particular areas of the brain.

Because the brain is so susceptible to slight chemical imbalances, the need for homeostasis is greater in the brain than in any other organ. Homeostasis in the brain is unique and depends on the functional capacity of the capillary networks that supply blood to the brain. The brain is able to absorb some substances (e.g. O_2 and steroid hormones), but not others, because the capillaries are especially adapted to create the **blood–brain barrier**. This restricts the entry of potentially harmful substances, whilst allowing the entry of nutrients and O_2. The brain capillaries have walls composed of endothelial cells joined by tight junctions (see Book 1, Section 2.8), which prevent the passage of specific substances (Figure 2.11).

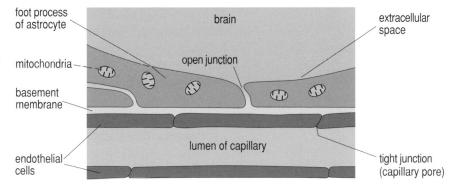

foot process of astrocyte

mitochondria

basement membrane

endothelial cells

brain

open junction

lumen of capillary

extracellular space

tight junction (capillary pore)

Figure 2.11 Cells of the blood–brain barrier. The endothelial cells of a capillary wall in the brain are shown surrounded by a basement membrane. The foot processes of astrocytes (containing numerous mitochondria) are shown in close contact with this basement membrane. Open junctions between the astrocyte end-feet provide direct access to the extracellular space around brain cells.

The capillaries are also surrounded by many cellular processes derived from astrocytes (called 'foot processes' because of their shape) which transfer metabolic products from the capillaries to the surrounding neurons and vice versa. Once fluid has entered the brain, it is known as cerebrospinal fluid (CSF) and flows through the brain into the fluid-filled channel at the centre of the spinal cord. You will have come across CSF in Book 2, Section 1.3.2. Note that CSF is not the same as blood plasma; it is a clear, colourless fluid with a high protein content, and has a constant pH which is not influenced by fluctuations in plasma pH. The astrocytes also pick up excess K^+ ions (produced during the transmission of electrical impulses in nerves) and any surplus neurotransmitters, which would harm the brain if allowed to accumulate, and deliver these, via their foot processes, to the capillaries, and so into the general circulation from where they are removed.

Because the blood–brain barrier is so well developed, many chemical substances that are essential for the brain, but which the brain cannot synthesize, e.g. glucose and amino acids, have to be transported across the barrier by active transport (see Book 1, Section 2.5.1). The mitochondria release the energy (ATP) required for active transport. The only substances that pass passively across the blood–brain barrier are lipid-soluble ones, such as O_2 and CO_2, as well as nicotine, caffeine, alcohol and heroin.

● Can you suggest why only lipid-soluble substances can passively cross the blood–brain barrier?

● Like all cell membranes, the membranes of the endothelial capillary cells are composed primarily of lipids, allowing lipid-soluble substances to pass through easily.

Water-soluble substances (such as Na^+, K^+ and Cl^- ions) are only able to cross the blood–brain barrier by energy-requiring active transport. Thus the blood–brain barrier tightly controls what enters the brain and contributes directly to this organ's homeostasis.

2.5.3 Circulation within the skin

The circulation of blood through the skin accounts for about 8% of the total cardiac output, but circulation to the skin can be increased dramatically, and rapidly, by up to 150-fold.

● Can you think of a physiological situation that causes a rapid increase in blood flow to the skin on the face?

● No? Shame on you! Don't you feel embarrassed? Blushing, especially in embarrassing social situations, is an example of sudden vasodilation allowing a rapid increase in blood flow to the skin, which produces a red flush on the cheeks of an embarrassed person.

● Why is it important to be able to increase the circulation through the skin so dramatically?

● To increase heat loss via the skin, aiding thermoregulation. Conversely, the ability to shut down blood flow through the skin will act to decrease heat loss, thus preserving body temperature.

The circulation within fingers, toes, palms and earlobes has special networks or 'shunts' which connect arterioles directly with venules, bypassing the capillaries. The shunts are situated close to the skin surface so that the blood may lose heat easily. These shunts have thick muscular walls which are richly supplied by nerves. When body temperature is increased, sympathetic activity to the skin decreases, causing vasodilation, particularly in the shunts. Overall, there is a decrease in peripheral resistance, leading to increased cardiac output and an increase in blood flow to the skin regions. Heat loss is increased as blood flow through the skin increases.

2.5.4 Skeletal muscle circulation

Under resting conditions, skeletal muscle receives about 15% of the total cardiac output. However, during exercise this can increase to 90%, and correspondingly O_2 consumption of exercising muscle can increase from 20% at rest to 90% of the body's total consumption. The increase in blood flow to the skeletal muscles (which is needed to sustain exercise) is brought about by vasodilation in the skeletal muscle capillary beds. A decrease in peripheral resistance and an increase in venous return from exercising muscles will in turn increase cardiac output.

2.5.5 Circulation to the kidney

Under normal conditions the renal arteries deliver about one quarter of the cardiac output to the kidneys; approximately 1200 ml of blood per minute enter the renal circulation. The kidneys are unique in that they have two very different capillary beds, each designed for a specific function. The primary role of the kidneys is to filter the blood, a process that occurs in the first capillary bed of the renal circulatory system, in the glomerulus. Blood enters the glomerular capillaries via *afferent* arterioles fed by the interlobular arteries (tributaries of the renal artery) and is drained from the glomerulus into the *efferent* arterioles.

● What is special about the arteriole arrangement of the glomerular capillary bed?

● Unlike any other capillary bed found in the body, the glomerular capillary bed drains into an *arteriole* and not a *venule*.

This unusual arteriolar arrangement is an important feature of the glomerulus. In the glomerulus, filtration occurs because of the extraordinarily high blood pressure generated within the glomerular capillary bed.

● Why is blood pressure important for filtration?

● Filtration is a direct consequence of the high blood pressure – it provides the energy to force the filtrate out of the blood and into the tubules.

The arterioles that feed and drain the glomerular capillary bed are high resistance vessels. The afferent arterioles have a larger diameter than the efferent arterioles and it is this arrangement that allows a high pressure to be maintained across the glomerular capillaries. In addition, the resistance of the afferent arterioles protects the glomerulus from large fluctuations in systemic blood pressure whilst the efferent arterioles reinforce the high glomerular pressure.

The second capillary bed of the renal circulation, which is a low pressure system like most other beds found in the body, is the peritubular bed, together with the vasa recta (Figure 1.8). These arise from the efferent arterioles draining the glomerulus and are intimately involved with the tubule, readily absorbing solutes and water from the tubular cells. The peritubular beds return their blood via the renal venules.

Summary of Section 2.5

1   The systemic circulation is the flow of blood around the body.

2   The coronary arteries provide the circulation for the heart muscle. Any disturbance in blood flow to cardiac muscle can jeopardize heart function and survival.

3   Blood flow to the brain is carefully controlled, thereby ensuring that there are always sufficient O_2 and nutrients available. The blood–brain barrier provides a protection against entry of most of the unwanted substances to the brain.

4   Circulation within the skin can be modified by shunting and this mechanism plays an important role in the control of body temperature.

5   Circulation to skeletal muscles can be increased greatly during exercise by the effects of vasodilation in the muscle capillary beds.

2.6 Abnormal cardiovascular function and its control by drugs

Worldwide, about 17 million people die every year from cardiovascular disease. In the UK alone, 39% of all deaths relate to complications associated with cardiovascular function. Many of these problems, as we will learn, can be significantly reduced by the use of prescribed drugs (Box 2.2), surgery or changes in lifestyle.

Box 2.2 The use of cardiac drugs

The choice and dosage of drug(s) used to treat a cardiovascular condition are crucial and depend on the nature of the condition and the person's medical history. The latter consideration may mean that a pre-existing illness excludes the use of a particular drug (called a **contra-indication** to the drugs usage). In addition to their direct therapeutic value, most cardiac drugs have *side-effects* (both acceptable and unacceptable) which may influence not only other heart functions but also the activities of other organs. Drugs act by binding to specific proteins (e.g. receptors, ion channels or enzymes) in the target structure; the specificity with which drugs bind to their target whilst ignoring closely related molecules, is central to drug performance and selection. However, since no drug acts with complete specificity, their use is a balanced compromise between benefits and side-effects. In practice, more than one drug is frequently prescribed to treat a cardiovascular condition. Since these drugs usually have to be continued indefinitely, initial treatment involves them being introduced in a stepwise manner at the lowest possible dosages, starting with the drugs least likely to produce side-effects.

Information about drugs used to alleviate cardiovascular malfunction can be found at the British National Formulary (BNF) website (see Further reading).

2.6.1 High blood pressure

High blood pressure is the major factor responsible for most of the deaths attributable to abnormal cardiovascular function. In the UK and USA, about 25% of all adults have abnormally high blood pressure. The normal systolic/diastolic blood pressure is about 100–160 mmHg over 60–90 mmHg. If the systolic reading repeatedly exceeds 160 mmHg, or the diastolic pressure is consistently above 95 mmHg, then the person may have high blood pressure or hypertension. However, these blood pressure values depend on age, obesity, inherited tendency to heart disease, and important lifestyle factors such as diet, smoking, exercise, emotional stress, etc.

● Can you think of one other important condition, which you have come across earlier in the course, where there is a potentially serious risk of developing high blood pressure?

● If a baby is born to a mother who was malnourished or who suffered a lack of oxygen whilst pregnant and if the baby was of a 'low birth weight to placental weight', then there is a significantly increased risk of developing hypertension in adult life (see Book 1, Section 1.5).

Blood pressure varies considerably throughout 24 hours and is greatly affected by emotional, mental and physical behaviour. For example, apprehension or sudden stress can cause a dramatic increase in blood pressure for brief periods. In fact, even being in hospital or visiting your doctor can elevate blood pressure – due to so-called 'white-coat hypertension'. This is why during a hospital consultation, sometimes a person's blood pressure is taken more than once, separated by a suitable period of time in order to obtain a more accurate measurement.

● Separate the factors influencing blood pressure mentioned above into those that can and those that can't be controlled or eliminated.

● Lifestyle factors such as diet, exercise, smoking and stress can all be changed (given the motivation and/or opportunity) to lower blood pressure. But inherited factors, such as tendency to heart disease, can't be eliminated. And then there's one factor that affects us all – the process of ageing. (Book 4, Chapter 1 has more details.)

● Why do you think it is important to lower the blood pressure of people who are hypertensive?

● Because high blood pressure can lead to a wide range of vascular complications, including coronary heart disease. (Other conditions include damage to the retina, kidney failure, cerebral haemorrhage and congestive heart failure. In the latter condition the performance of the heart gradually decreases so that there is congestion of blood in the venous system (see Section 2.6.3). Maternal high blood pressure during pregnancy (**eclampsia**) can also seriously compromise the normal functioning of the placenta, which can significantly affect fetal development.)

Hypertension is associated with an increase in peripheral resistance related to a reduction in elasticity of the blood vessels and a gradual narrowing of their internal diameters. This results in the heart having to work harder to overcome the increase in resistance. Frequently, the muscle of the left ventricle enlarges significantly (*hypertrophies*), thereby maintaining an appropriate cardiac output. Under such conditions the functional capacity of myocardial cells may begin to deteriorate markedly, as in heart failure (see below).

Non-drug therapies to reduce high blood pressure include stopping smoking, weight reduction, salt restriction, increased aerobic exercise and emotional stress management. Where these therapies produce no consistent beneficial reduction in blood pressure, anti-hypertensive drugs may need to be prescribed. The first-line of drugs used in the treatment of hypertension are the β-**adrenoceptor antagonists** (commonly called *beta-blockers*), the **diuretics** and the **vasodilators** (Figure 2.12).

Figure 2.12 Drug treatment of hypertension. See text for details.

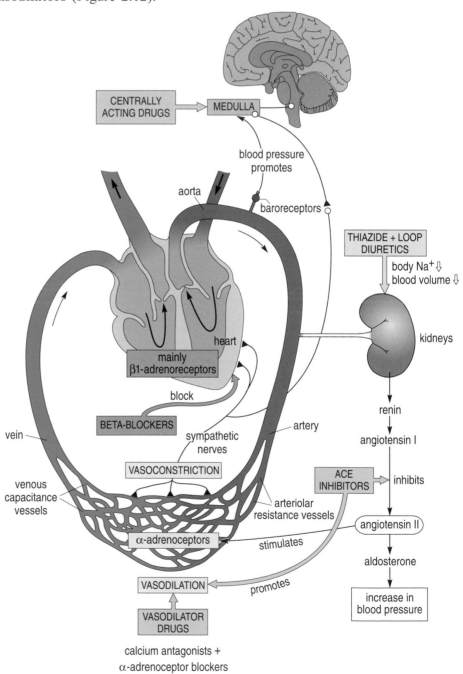

You may already have heard of 'beta-blockers', but what do they block? The sympathetic nerves ending in the heart use the neurotransmitter noradrenalin to stimulate specific receptors on the surface of cardiac cells called **adrenoceptors**. Adrenoceptors in the body are divided into alpha (α1 and α2) and beta (β1 and β2) subtypes. α1-, α2- and β2-adrenoceptors are found in bronchial, vascular, intestinal and uterine smooth muscle. Both β1- and β2-adrenoceptors are present in heart muscle, but the β1 subtype greatly outnumbers the β2 subtype in a ratio of about 3 : 1. Thus beta-blockers, in the context of heart function, are specific receptor antagonists that 'block' the action of cardiac β1-adrenoceptors.

● Name a circulating chemical in the blood that also stimulates β1-adrenoceptors.

● The hormone adrenalin (see Section 2.4.6 above).

● Can you foresee any side-effects of drugs aimed at β2-adrenoceptors in bronchial, vascular, intestinal and uterine smooth muscle?

● Yes. They may also influence heart function by affecting cardiac β2-adrenoceptors.

The commonly prescribed beta-blockers *metoprolol* and *atenolol* produce a fall in blood pressure by decreasing cardiac output. They reduce the stimulating action of noradrenergic nerve input to the heart by selectively blocking the β1-adrenoceptors. The result is that heart rate, force of heart muscle contraction and cardiac output are all reduced.

● What mechanism might underlie this reduction in blood pressure?

● Since the blockade of β1-adrenoceptors reduces cardiac output, then for a given peripheral resistance, blood pressure will be reduced because cardiac output × total peripheral resistance = mean arterial blood pressure.

Unwanted effects of β1-adrenoceptor-blockers include fatigue due to reduced cardiac output and therefore a reduction in blood supply to the muscles, especially during physical exercise. Unfortunately, β1-blockers can also affect β2-adrenoceptors, causing a loss of β2-adrenoceptor-mediated vasodilation in the extremities which results in cold hands, feet and nose. In addition, the use of beta-blockers can cause an unwanted degree of cardiac failure in patients already suffering heart disease – in such patients the use of beta-blocking drugs is therefore contra-indicated.

In Chapter 1 you learnt about the function of the kidney and diuretic drugs that promote an increase in water excretion from the body. There are two important classes of diuretic drugs: **thiazide diuretics** which act on the distal tubules of the kidneys and **loop diuretics** which act on the 'thick ascending loop of Henle' in the kidney. The effect of both diuretic drugs is to produce a reduction in circulating blood volume (due to water loss) and a related decrease in venous return with a consequent reduction in blood pressure.

Vasodilator drugs reduce blood pressure by decreasing vasoconstriction and hence peripheral resistance (Figure 2.12). The tone of vascular smooth muscle cells (Book 2, Section 4.7.1) is determined by the concentration of Ca^{2+} ions in their cytosol. **Calcium antagonists,** for example *nifedipine*, reduce the tone of arterioles by blocking the entry of Ca^{2+} into cells through calcium ion channels, which causes a relaxation of the smooth muscle (Figure 2.13). This calcium blockade therefore produces vasodilation and a fall in peripheral resistance which leads to a reduction in blood pressure. Vasodilation can also be produced by drugs (e.g. *prazosin*) that selectively block $\alpha 1$-adrenoceptors; these adrenoceptors are normally involved in noradrenalin-mediated vasoconstriction (Figure 2.13). The potential side-effects of using $\alpha 1$-adrenoceptor blockers include flushing and headache.

Some cases of hypertension are due to an overstimulation of the **renin–angiotensin–aldosterone system** (see Section 1.5 of this book). In this system, angiotensin I is converted by a specific enzyme into angiotensin II, which promotes the secretion of aldosterone from the adrenal cortex. Angiotensin II is also a potent vasoconstrictor and aldosterone is directly involved in salt and water retention – the production of both substances leads to a significant increase in blood pressure. Inhibitors of the *angiotensin I converting enzyme* are called **ACE inhibitors.** Such inhibitors (e.g. *captopril* and *enalapril*) are therefore very important antihypertensive drugs (Figure 2.12). The use of ACE inhibitors in hypertensive patients results in marked decreases in vascular resistance (due to vasodilation) and blood volume (due to sodium and water excretion); both effects can produce a significant lowering of blood pressure. Aldosterone production can be inhibited directly by the antagonist *spironolactone*. One important side-effect of these vasodilators is that they can produce very low blood pressure (**hypotension**), possibly leading to renal failure (Section 1.4). An alternative is to block the action of angiotensin II by using drugs called **angiotensin II receptor antagonists** (e.g. *losartan*). However, as with ACE inhibitors, such drugs should not be used in people with impaired renal function, as further kidney damage may ensue.

There are also specific drugs that act directly on the central nervous system (called **centrally acting drugs**), such as *clonidine* and *α-methyldopa*, which affect cardioregulatory regions of the medulla in the brain to produce a reduction in the activity of sympathetic nerves innervating the heart. This decreases the force of heart muscle contraction, cardiac output and therefore blood pressure (Figure 2.12). Since adequate O_2 uptake and transport is severely affected, this limits the extent to which an individual can participate in physical exercise. The use of these drugs is frequently accompanied by headache, nausea and sedation. A specific feature of drug use is illustrated by clonidine: if the drug is suddenly withdrawn, the patient's hypertensive symptoms swiftly reappear via a mechanism called *rebound hypertension*.

An interesting perspective on hypertension is given by Brian Mills, an industrial engineering manager from Australia, who in 1989 wrote a poignant article entitled '80,000 pills: A personal history of hypertension'. Brian was diagnosed with hypertension in 1950 at the age of 21 years. At 24 years of age Brian's blood pressure was 240/140 mmHg and kidney damage was already apparent. To combat his hypertension, Brian took on average about six tablets a day between 1950 and 1989. Unfortunately, such was the specificity of the drugs available during the 1950s–1970s, that the great majority of the tablets prescribed during this period were to counteract the side-effects of the medication taken to reduce the initial hypertension.

2.6.2 Coronary heart disease

Next to cancer, coronary heart disease (CHD) is the most common cause of death in the UK, the USA and most other industrialized societies. In the UK, 21% of all people die from CHD and its complications. The leading cause of CHD is atherosclerosis, which is characterized by deposits of lipid, cellular debris and calcium salts on the inside of arterial walls. As these deposits (called **plaques**) build up, they narrow the artery and reduce the elasticity of the artery walls, eventually reducing blood flow through the vessel (see Figure 3.11 in Book 1). (*Note:* do not confuse atherosclerotic plaques with dental plaques; see Box 4.1 in Book 1).

Although the underlying cause of atherosclerosis is not known, the condition appears to begin with damage to the endothelial cells lining the arteries. Once the damage has begun, the endothelial cells proliferate and lipids (e.g. cholesterol), cell debris and calcium salts are deposited to form the basis of the blockage. Epidemiological evidence has identified several factors that increase the progression of atherosclerosis; two of these factors are cigarette smoking and a sedentary lifestyle.

● Having learnt about nutrition in Book 1, what other factor do you think contributes to the development of atherosclerosis?

● A diet containing large quantities of saturated fatty acids is likely to contribute to the development of atherosclerosis (Section 3.4 of Book 1).

Drugs such as *statins* reduce high cholesterol levels in the blood (**hypercholesterolaemia**), and are therefore very important in the fight against CHD. Prescribed drugs, for example *simvastatin* and *pravastatin*, significantly lower blood low density lipoprotein (LDL, 'bad'), cholesterol and triacylglycerol levels by blocking production of the specific liver enzymes that synthesize cholesterol and lipids. Help is at hand from a new genetically engineered protein called apoA-1 Milano (Nissen et al., 2003). This drug significantly lowers blood plasma LDL levels and appears to unclog the narrowed arteries of CHD patients 'like a drain cleaner'.

● Can you suggest other factors which may underlie the progression of CHD?

● Other factors linked to CHD include hypertension and stress. Diabetes, age, gender, and, in particular, genetics are also contributory factors.

The typical consequence of CHD caused by atherosclerosis, is a reduced supply of O_2 and nutrients to the heart muscle, due to decreased blood flow through the narrowed coronary arterial circulation (**coronary artery disease**). This is termed myocardial ischaemia (defined as lack of blood flow; from the Greek *iskhein*, 'to hold back' and *haima*, 'blood'), or local ischaemia if it relates to an isolated area of heart muscle. If the ischaemia is mild, then usually enough O_2 and nutrients are delivered to the heart tissue to just sustain normal metabolic activity. However, if the heart rate is increased (e.g. during exercise, over-eating, or stress) and the O_2 supply becomes insufficient, this may lead to a chest pain called angina pectoris (see Book 1, Section 3.4.1; also Figure 2.13 and Case Report 2.1 in this chapter).

● If the O_2 supply to heart muscle is insufficient, what do you think the basic aim of treatment in angina might be?

● To reduce the oxygen demand, by decreasing the work the heart has to do and therefore its level of muscular activity.

The first line of drugs used to treat angina are nitrate compounds called **nitrovasodilators** –for example, *glyceryl trinitrate*, which is rapidly absorbed into the blood circulation when placed under the tongue (*sublingually*) or inside the lip *(buccally)*. Nitrovasodilators are used to terminate acute angina attacks (frequently within minutes of administration) and can be combined with other drugs to provide sustained chronic treatment. The absorbed drugs circulate in the blood and react with compounds in vascular smooth muscle cells to produce nitric oxide (NO), a rapid potent vasodilator. At therapeutic doses, nitrovasodilators act primarily to dilate veins, thus causing a retention of blood in the peripheral circulation. This reduces venous return, resulting in a decrease in ventricular end-diastolic volume and a reduction in the active distension of the heart wall. The oxygen demand of the heart is thereby decreased, relieving the chest pain (Figure 2.13). A common side-effect of nitrovasodilators is that patients frequently suffer from headaches due to dilation of cerebral blood vessels. The reduction in blood pressure can cause fainting – an additional side-effect. Beta-blockers and calcium antagonists are also used to treat angina by reducing blood pressure and peripheral resistance (Section 2.6.1 and Figure 2.13). However, with repeated and prolonged use, people can become tolerant to anti-anginal drugs.

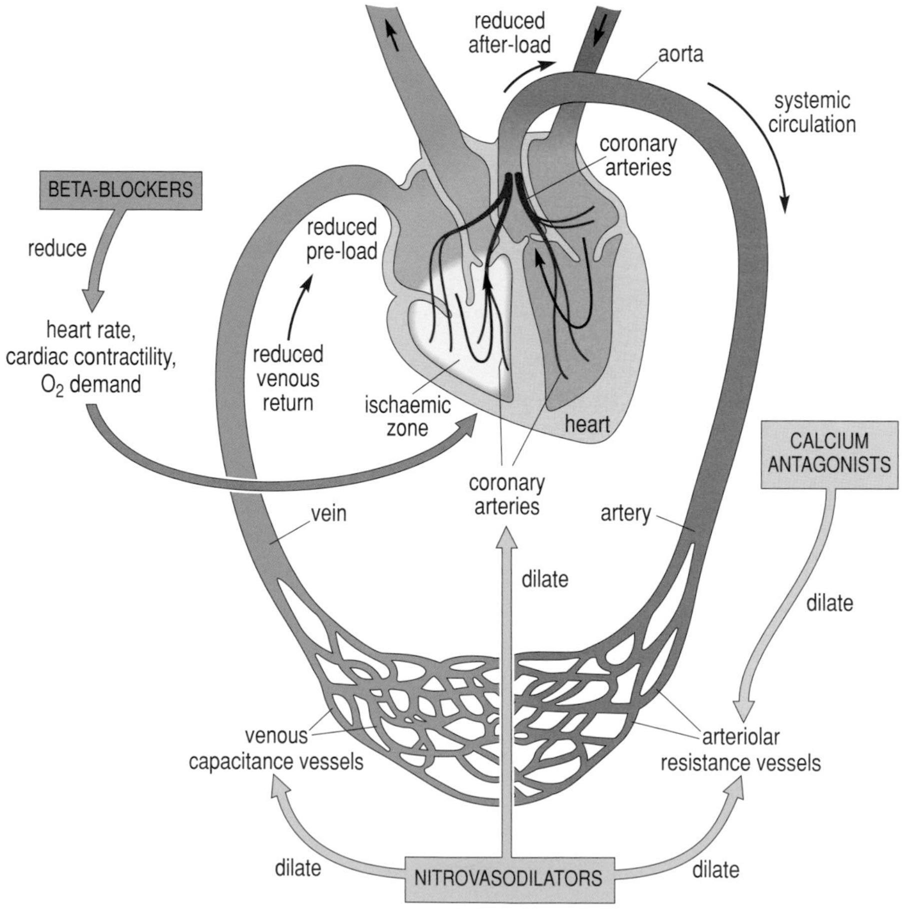

Figure 2.13 Diagram showing the use of drugs to treat angina pectoris and their cardiovascular effects.(The terms pre-load and after-load are explained in Section 2.6.3.)

Case Report 2.1 Coronary heart disease

Mohammed is a 49-year-old father of four children, who works as a taxi driver in London. He has been referred by his GP to the rapid-access chest pain clinic, with a two-year history of 'chest pain on exertion' which has markedly worsened over the last three months. Mohammed has a family history of coronary heart disease and is known to have a raised blood cholesterol level for which he is taking *simvastatin* (20 mg daily). He smokes 15–20 cigarettes daily. Mohammed has no history of diabetes mellitus and he explained that the frequency of chest pain was variable and described it as a pressure across his chest accompanied by feeling sweaty and a shortness of breath. He said the pain did not radiate into his arms and it disappeared when he rested. Mohammed's blood pressure was recorded as 130/80 mmHg and his pulse was measured as 80 beats per minute.

An exercise stress test was then carried out. This entailed recording an electrocardiogram (ECG) at regular intervals while Mohammed was resting and then while he was exercising on a treadmill. At rest his ECG was normal (Figure 2.14a). After 3 minutes of exercise, Mohammed started to develop slight chest tightness. After 6 minutes, his chest pain was so intense and he was so short of breath, that the test was stopped at 6 minutes 45 seconds. The chest pain eased within 3 minutes of resting. The ECG recorded at 6 minutes 42 seconds revealed developing

ischaemia, as diagnosed by characteristic horizontal/downward-sloping depression of the ST segments and inverted T waves (Figure 2.14b) – typical indicators that the ventricular muscles are not getting a sufficient supply of oxygen. The results of the tests indicated that his pain was almost certainly cardiac in origin and Mohammed was informed that he needed to have an angiogram. He was strongly advised to stop smoking and to attend the smoking cessation clinic. He was told that if he got chest pains that did not disappear on rest and lasted more than 15 minutes, he should ring for an ambulance straightaway, as this could be symptomatic of a heart attack. His cholesterol-lowering drug (*simvastatin*) was increased to 40 mg daily, and he was prescribed a beta-blocker (*bisoprolol*, 2.5 mg daily) and *aspirin* (75 mg daily) to prevent platelet aggregation and blood clot formation. Mohammed was naturally concerned about the results of his tests and the implications for his work. He was reassured that he would be able to continue to drive and carry out his normal physical activities. However, he was told that if he developed chest pains while driving, he was to stop the vehicle. Before going home to his family, Mohammed was given a selection of leaflets reiterating the advice given to him and clearly explaining his medication and the tests he had undergone.

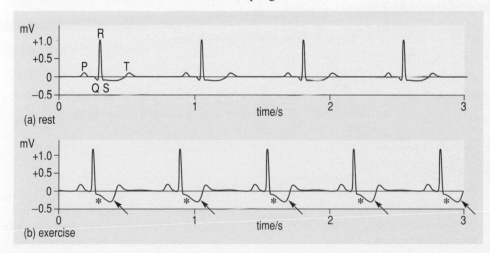

Figure 2.14 Mohammed's ECG at rest (a) and during exercise (b). The ECG recorded during rest is normal (compare with Figure 2.5a and b). The ECG recorded 6 minutes 42 seconds after the start of exercise shows characteristic signs of ischaemia, i.e. horizontal/downward-sloping depression of the ST segments (asterisks) and inverted T waves (arrowed).

When angina either increases in severity, has a rapid onset during very mild exercise, or does not respond to chemical intervention, **balloon angioplasty** may have to be performed. An **angiogram** of the heart (where dyes are injected into the coronary circulation to reveal how blood travels through the coronary circulation) will indicate which coronary vessels are becoming occluded and the extent of the blockage(s). Subsequently, the affected coronary arteries can be stretched open by expanding a special balloon in the 'furred-up' (atherosclerotic) segments. Small hollow metal tubes, called **stents**, can also be inserted inside these regions to keep such arteries 'open', which can help to restore the blood supply to the tissues without having to resort to coronary artery bypass surgery. In the latter surgical procedure, arteries (or veins) from different parts of the body are grafted to the affected coronary arteries to bypass the atherosclerotic segment(s). Patients who have undergone bypass surgery show a dramatic post-operative relief from the symptoms of angina. In some cases of coronary heart disease, an acute *myocardial infarction* (*MI*) may occur. An **infarct** consists of a central area of dead cells surrounded by tissue that is ischaemic (lacking in oxygen and nutrient supply) but not yet dead. In acute MI, the O_2 supply is so severely reduced that the region of the heart muscle affected may die (**necrosis**) within an hour or so and as a result, the activity of the whole heart may be severely threatened.

●   What do you think an acute myocardial infarction is commonly called?

●   A 'heart attack'.

Evidence of heart muscle injury and necrosis following a heart attack can be readily diagnosed in ECG traces that display abnormal QRS waveforms and ST segments (Section 2.4.2). In addition to unusual ECG traces, the necrotic muscle tissue releases certain proteins (e.g. troponin) into the blood that can be detected biochemically.

There are two common ways whereby atherosclerosis can cause a heart attack. Firstly, an atherosclerotic coronary artery may become totally occluded preventing the passage of oxygenated blood downstream – this may take many years to occur. Alternatively, an atherosclerotic coronary blood vessel can become completely blocked by a freely moving blood clot formed elsewhere in the coronary circulation – this is explained in more detail in Section 2.6.5 (below).

It is now widely accepted that the relationship between hypertension and CHD is intertwined, having no discernible starting point and a gradual development. Unfortunately, hypertension and CHD perpetuate themselves: elevated blood pressure predisposes blood vessel walls to develop atherosclerosis, which further elevates blood pressure. A vicious cycle, gradually leading to compromised cardiovascular function, has begun…

Research has shown, however, that 'a problem shared is another heart attack halved.' Researchers at Manchester Royal Infirmary studied the lifestyles of 600 people (mean age 60 years; 75% of whom were male) for the first year after they had suffered a heart attack (Dickens et al., 2004). It was found that patients who were able to confide in a close companion were half as likely to suffer subsequent heart attacks as those without such a confidant(e). Moreover, another study (Allen et al., 2003) indicated that the ownership of domestic pets significantly reduced blood pressure and cardiovascular stress. (The subject of psychological stress and the cardiovascular system is considered further in Book 4.)

2.6.3 Congestive heart failure

The underlying problem in congestive heart failure (CHF) is that cardiac output is so abnormally low that it is unable to produce sufficient blood pressure in tissues of the body. This can be the result of progressive hypertension or CHD and can lead to a variety of symptoms such as fatigue, breathlessness and fluid retention in the peripheral circulation. In CHF, the heart is required to work harder to achieve a sufficient output. However, when cardiac output declines there is a gradual build-up of fluid, which causes an increase in the return of venous blood to the heart (termed **pre-load**). In addition, the reduced flow through the renal system activates the renin–angiotensin–aldosterone system, causing significant peripheral vasoconstriction. This causes an increase in arteriolar resistance against which the heart has to pump blood (termed **after-load**). An increase in both pre-load and after-load provides a failing heart with a significant uphill challenge. A reduction in both of these loading factors is required so that effective cardiac output can be restored.

● What is the basic aim in the treatment of CHF?

● The aim is to reduce circulating blood volume and to decrease the force of heart muscle contraction, thereby reducing the work the heart has to do.

Treatment for CHF usually starts with diuretics, which as explained in Section 1.5, increase the excretion of Na^+ ions and water, and by reducing the circulatory volume decreases the oedema and 'pre-load'. In addition, vasodilators can be prescribed. The ACE inhibitors *(captopril* and *enalapril)* produce both venous and arteriolar dilation, thus reducing vascular resistance by preventing the increase in angiotensin II (a vasoconstrictor) and the ensuing production of aldosterone. ACE inhibitors increase Na^+ and water excretion, which reduces blood volume and 'pre-load'. As cardiac output increases, blood pressure rises and renal blood flow increases and heart function improves.

In severe heart failure, if a combination of diuretics and ACE inhibitors fails to provide adequate response, then an **inotropic drug** is required (inotropy was introduced in Section 2.4.5). Inotropes, such as *digoxin* (extracted from leaves of the foxglove, *Digitalis*), increase the force of cardiac muscle contraction by indirectly increasing the intracellular concentration of Ca^{2+} ions in heart cells. Digoxin is an inotropic drug. Other inotropic substances, *dobutamine* and *dopamine*, act via directly stimulating the β1-adrenoceptors in heart muscle. The effect of both drugs is to increase cardiac output and blood pressure, which improves the flow of blood through the tissues and greatly reduces the 'load' on the heart.

2.6.4 Cardiac arrhythmias

As we have learnt, the rhythm of the heart is normally determined by the regular electrical activity of pacemaker cells in the SAN (see Section 2.4.1). Abnormal heart rhythms are termed cardiac arrhythmias and can be identified by characteristic alterations to the normal ECG trace (a brief revision of Section 2.4.2, especially Figure 2.6 may be useful here). Cardiac arrhythmias can arise from: unusual sites generating electrical activity (so-called *ectopic foci*), alterations in SAN pacemaker activity, or by interference with the conduction of action potentials through the heart. We now consider three examples of cardiac arrhythmia.

Atrial fibrillation

Atrial fibrillation (AF) can accompany several heart conditions (e.g. coronary heart disease and hypertension) and is characterized by rapid uncoordinated depolarizations of the atria, which quiver (or 'fibrillate') with heart rates increasing up to 150–180 beats per minute. Chronic AF commonly affects 5% of (1 in 20) people over the age of 65.

In atrial fibrillation, impulses reach the AV node erratically, resulting in a ventricular rhythm that is not synchronized with atrial contraction. Consequently, cardiac output may be inadequate to generate sufficient blood pressure for adequate tissue perfusion. This may precipitate heart failure. The ECG in AF is characterized by 'fibrillating' atrial activity instead of normal, clearly defined P wave depolarizations; this is usually accompanied by irregularly timed QRS complexes (see Figure 2.6c).

The aim of treatment in AF is to control ventricular rate by re-establishing the normal atrial 'pacemaker' (or sinus) rhythm (see Section 2.4.1). The process of re-establishing the normal sinus rhythm is called **cardioversion**. Cardioversion can be accomplished by drugs, or by an electrical shock to 'defibrillate' the heart. The latter procedure is achieved by applying timed electrical currents across the person's chest. The aim is to interrupt the chaotic electrical activity by simultaneously depolarizing most or all of the myocardial cells; this hopefully allows the normal electrical pacemaker activity, or sinus rhythm, of the heart to be resumed. Approximately 70–90% of patients with chronic AF can be converted to sinus rhythm by electrical cardioversion. However, 60–75% of patients cardioverted successfully using applied electrical currents will revert to AF within one year.

Cardioversion can also be achieved pharmacologically by drugs (e.g. *verapamil*) that act by increasing the refractory period of the AVN, bundle of His and Purkinje fibres. Although each of these drugs uses a different physiological mechanism to achieve a prolongation of the AVN delay (Section 2.4.1), the overall effect is a reduction in atrioventricular coupling so that the frequency of impulse transmission from the atria to the ventricles is significantly reduced.

- Verapamil acts by inhibiting the entry of Ca^{2+} ions into cardiac cells. If a patient who is already using beta-blockers for severe hypertension starts to get atrial fibrillation, would you prescribe verapamil?

- No. The use of beta-blockers contra-indicates the use of verapamil. A person using beta-blockers will already have a slow heart rate and lowered force of cardiac muscle contraction. Taking verapamil in addition to beta-blockers would further reduce heart rate and the force of muscle contraction, leading to possible 'cardiovascular collapse' and death.

Of note is that patients with AF are immediately given *warfarin* to reduce the likelihood of blood clotting. Prolonged AF causes a stagnation of blood flow in the heart chambers that can lead to small blood clots forming, with the possibility of subsequent heart attack or cerebral stroke (see below).

Ventricular fibrillation

The onset of **ventricular fibrillation** (VF) is sudden and the heart can completely stop (cardiac arrest). It is a true emergency, as there is no cardiac output. VF can result from coronary heart disease, imbalance of electrolytes (dissolved ions) in the body (e.g. low levels of K^+ or Mg^{2+} ions), or as a result of other cardiac arrhythmias. The first sign of VF is usually unconsciousness due to the brain not getting enough oxygen. Immediate intervention is necessary, either through cardiopulmonary resuscitation (CPR) or by restoring the normal electrical activity of the heart (see Further Reading).

As there is no cardiac output in VF, there is no effective blood pressure. The ventricles receive the normal rhythmic 'pacemaker' potential, but the ventricular musculature exhibits uncoordinated contractions, because multiple impulses travel erratically around the ventricles, with heart rates reaching over 300 beats per minute. The ECG waveform and baseline in VF are both highly irregular, lacking the characteristic profiles of distinct QRS complexes (Figure 2.6d).

Another type of conduction abnormality is **atrioventricular (AV) heart block**, which arises from defects in the electrical conducting systems (AV node and bundle of His) between the atria and ventricles of the heart. Symptoms often include dizziness and fainting as a result of reduced cardiac function and output. There are several defined types of *partial AV heart block* in which impulses between the atria and ventricles are blocked to varying degrees. First-degree AV block results from a delay in conduction through the AV node, which increases the interval between the atrial and ventricular contractions. In second-degree AV block (called *Mobitz Type II*) only a proportion of the atrial impulses get through to the ventricles. For example, only every second atrial impulse might be transferred to the ventricle (called '2 : 1 AV block').

● What do you think '3 : 2 AV block' represents?

● That for every three atrial impulses only two ventricular beats occur.

By comparison, in *complete AV heart block*, although atria and ventricles have regular contraction rates, they are unrelated due to a damaged atrioventricular node – as might occur after myocardial infarction. In an ECG of a patient with complete AV heart block, the P waves exhibit a normal rhythm; however, the QRS and T waves occur at a much slower regular rate that is completely independent of the P wave rhythm (Figure 2.6e). So although atria beat regularly, the ventricles are not stimulated and consequently generate their own much slower intrinsic rate of about 35 beats per minute. Complete AV heart block can be effectively treated by surgically implanted electrical pacemakers. These miniature computerized devices monitor heart function over a wide range of physiological activities and can rectify abnormal cardiac rhythms by applying suitable electrical currents to the heart.

2.6.5 Problems of the circulation

We consider next some of the main problems associated with abnormal circulation of blood through the cardiovascular system:

* chronic arterial swelling;
* blockage of blood vessels;
* poor circulation;
* circulatory shock, which includes the loss of blood and low blood pressure.

Chronic arterial swelling

As we have learnt, blood is under high pressure as it is pulsed through the aorta. At weakened places, for example that might occur after a crush injury to the abdomen during a car accident, the aortic wall may swell or bulge outwards like a balloon in response to the blood pressure. The bulge is called an 'aneurysm' and most aortic aneurysms occur in the abdominal portion of the aorta. Apart from injury, the main cause of aneurysms throughout the body is hypertension resulting from gradual arterial atherosclerosis (Figure 2.15). Aortic aneurysms are potentially fatal and are repaired by surgically 'by-passing' the affected segment with graft tissue or by placing a more rigid cuff, similar to a stent, around the aneurysm.

Although much less serious, swelling can also occur to the walls of veins. In the condition called *haemorrhoids* (or *piles*), abnormal blood pressure or flow in the veins around the anal canal (for example, that might occur during pregnancy or as a result of constipation) may cause the vessel walls to 'balloon' at several places (Book 1, Section 3.3.3). Swelling may occur either inside or outside the anal sphincter. Haemorrhoids are common. By the age of 50, up to half the population of the UK will have suffered from them. A comforting thought is that haemorrhoids usually get better gradually without the need for medical attention.

Blockages in blood vessels

Blockages in blood vessels are very serious events. Not only can large atherosclerotic plaques partially or completely block blood vessels, but they can also break through the weakened endothelial lining, exposing the passing blood to the collagen in the underlying connective tissue (Figure 2.15a and b). When blood platelets come into contact with collagen, a clot, called a *thrombus* (Greek: *thrombolos* meaning 'lump') can be formed by coagulation of the blood (Figure 2.15c). The principal stages underlying clot formation and dissolution are shown in Figure 2.16 and described below. The thrombus may enlarge to block the blood vessel, as in myocardial infarction, or it may shear away from its anchor point to form a mobile *embolus* (Greek: *embolos* meaning 'stopper'; pl. *emboli*). A freely moving embolus may partially or completely 'plug' a smaller vessel as it flows downstream (Figure 2.15c and d). Thus through **thromboembolism**, atherosclerosis can result in a gradual or sudden blockage of any blood vessel, be it in the heart (*heart attack*; Section 2.6.2), or lungs (*pulmonary embolism*) or brain (*cerebral stroke*) – all potentially life-threatening events.

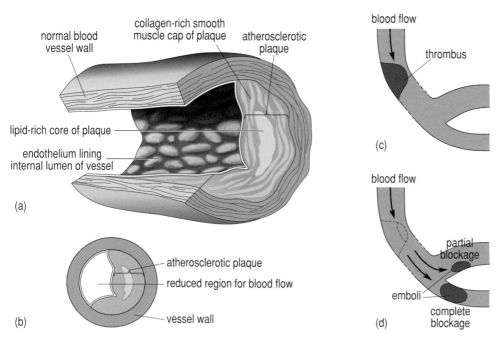

Figure 2.15 Atherosclerotic plaques. (a) Plaque structure and its relationship to the blood vessel in which it is situated. (b) Cross-section through the blood vessel showing the extent of obstruction to blood flow. (c, d) A thrombus can shear off into emboli (blood clots) that partially or fully block downstream blood vessels.

Cerebral stroke (Case Report 2.2) is the third most common cause of death in the Western world; indeed, over 100 000 people a year suffer a stroke in the UK. Stroke (also called 'cerebrovascular accident' or 'brain attack') accounts for about 12% of deaths from all causes in England and Wales. It often occurs in individuals who are over 65 years of age, but it may affect anyone, even those in the prime of life. In fact, each year over 10 000 people of working age in the UK have a stroke and a large proportion of these will subsequently experience severe disability. Moreover, about 10% of people who have had a stroke will have another one in the first year after the initial attack.

A stroke occurs when the flow of blood to a part of the brain is stopped for a period of time – the right or the left cerebral hemisphere may be affected. If a thromboembolism stops the flow of blood to a particular brain region for a sufficiently long period of time, the nerve cells in that area begin to die. This is referred to as infarction because it consists of a central area of dead cells surrounded by brain tissue that is ischaemic but not yet dead. *Haemorrhagic stroke* occurs when an aneurysm in a cerebral blood vessel ruptures and causes an internal bleed inside the brain. The most common symptom of a cerebral stroke is paralysis.

● If a person has a stroke in the left cerebral hemisphere which side of the body is likely to be affected?

● The right side. Stroke-related damage is most likely to affect the opposite (contralateral) side of the body to the side of the brain in which the stroke occurred.

The clinical signs associated with a stroke vary according to the region of the brain affected, and the type and severity of stroke, but can typically include:

• a sudden weakness or loss of feeling of the face, arm or leg on one side of the body;
• a sudden loss of vision, particularly in one eye;
• difficulty with or in understanding speech;
• unexplained dizziness,
• unsteadiness or sudden falls;
• inability to swallow;
• possible incontinence.

Predisposing factors associated with stroke include high blood pressure, smoking, heart disease, obesity, high cholesterol, diabetes and stress.

A brain scan is essential for a potential stroke-patient so that the location and extent of possible brain areas affected can be identified and possible susceptible regions of continued internal bleeding in the brain defined. For the best chances of recovery following a stroke, medical intervention needs to be started immediately, since the rate of recovery is most rapid in the first weeks and months after the incident. Subsequent personal and physical rehabilitation of the stroke patient also need to be commenced as soon as possible.

The formation of haematoma (blood clots), not only in the brain but throughout the body, can result from circulatory stagnation, damage to the internal surface of a blood vessel or in response to the loss of blood (haemorrhage) following an external injury that ruptures a vessel. The process that stops bleeding is called **haemostasis** (not to be confused with homeostasis).

A cut or torn blood vessel immediately constricts as a result of both an inherent vascular response to injury and sympathetically induced vasoconstriction. This vasoconstriction (termed **vascular spasm**) slows blood flow and minimizes blood loss until haemostatic processes take over. Platelets do not normally adhere to the smooth endothelial surface of blood vessels. However, following vessel wall injury, platelets are activated by the underlying collagen to aggregate into a **platelet plug** (Figure 2.16a) at the site of the defect, thereby sealing the vessel.

Haemostasis involves other specific blood-borne factors that are involved in a complex chain of biochemical reactions, called the **'clotting cascade'**. This haemostatic pathway ultimately converts fibrinogen into a stabilized net-like fibrin mesh (Figure 2.16a). Erythrocytes and additional platelets become entangled in the fibrin net, and together they promote the formation of a fibrous scar at the vessel defect, as well as being involved in the subsequent repair to the vessel wall. The clot, which is no longer needed, is slowly dissolved by a **fibrinolytic** (fibrin-splitting) enzyme called *plasmin*, and the products of clot dissolution are removed by scavenging phagocytic white cells (Figure 2.16b). To prevent clot formation at inappropriate sites in the vasculature, plasmin is constantly produced at low concentrations. The reverse of the latter situation, where clots fail to form in damaged vessels, leading to uncontrolled blood loss, is called **haemophilia** and primarily results from a genetic deficiency to produce one of the factors involved in the 'clotting cascade'.

The subsequent processes that are involved in wound healing have been presented in Book 1, Section 2.10.

As can be seen in Figure 2.16, there are strategic points in the biochemical pathways underlying haemostasis where clot formation can be influenced by drugs. For example:
• **antiplatelet** drugs (e.g. *aspirin*) can inhibit platelet aggregation;
• **anticoagulants** such as *warfarin* and *heparin* can inhibit the conversion of prothrombin to thrombin and so prevent the conversion of fibrinogen to fibrin; and
• **thrombolytics** ('clot-busting' fibrinolytic drugs), for example *streptokinase*, can be used to rapidly dissolve clots by activating the conversion of plasminogen to plasmin.

● When might 'clot-busting' drugs be of use?

● These are the first line of drugs used in the prevention of blood clot formation and treatment of possible thromboembolism occurring as a result of cardiovascular conditions such as CHD, heart attack[1], atrial fibrillation, deep vein thrombosis (see below), or after surgical operations (for example, artificial heart valve insertion or joint replacement).

The main side-effects of anticoagulants are bleeding and allergic reactions, whilst those of thrombolytic drugs are nausea, vomiting and possible bleeding.

[1] Of major significance in the UK is that guidelines – The National Health Service Framework for Coronary Heart Disease – were introduced in 2000, which officially recommend that thrombolytics are given within 1 hour, preferably as soon as possible, of a likely heart attack victim calling for medical assistance – this is termed 'the call-to-needle time standard'. National guidelines are currently being drawn-up for the whole patient care pathway, in order to standardize the treatments given to people presenting with cardiac emergencies.

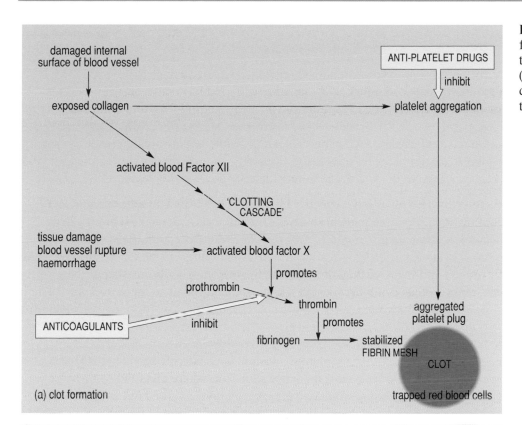

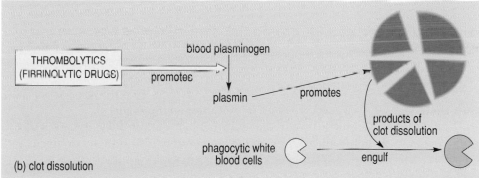

Figure 2.16 (a) Blood clot formation and the effects of therapeutic drugs on this process. (b) The sequence of clot dissolution and the action of thrombolytics (fibrinolytic drugs).

Case Report 2.2 Cerebral stroke

Gladys is a 72-year-old widow who has three grown-up children, two of whom live locally, and several grandchildren. She is a retired midwife, of Afro-Caribbean origin, and has lived in the UK for many years. Although Gladys is generally in good health and has never smoked, she has diabetes (controlled with tablets) and is overweight.

One Sunday morning in church, Gladys suddenly collapsed. The ambulance was called and the paramedics found her to be conscious, but she had difficulty in speaking and an obvious weakness in her right arm and leg. She was assessed and examined in the local Accident and Emergency Department ('A&E'). Gladys had high blood pressure, but her other 'vital signs' were within normal limits. (Vital signs are physical indicators of life, e.g. pulse, breathing, temperature.) Blood was taken for analysis and an intravenous infusion of saline started. Gladys was not allowed food orally as the staff suspected that she had experienced a stroke and there was a possibility that her swallowing reflex might have been affected. The A&E doctor referred Gladys to the consultant who examined her and explained that it was likely that she had sustained a cerebral stroke. He informed her that she would have an MRI scan as soon as possible to confirm the diagnosis and show whether the stroke had been caused by a clot (which was most likely) or by a haemorrhage within the brain.

The following morning, Gladys had an MRI scan which showed an area of acute infarction in her left temporal lobe. The stroke coordinator (a specialist nurse) explained this result to Gladys and her daughter, saying that the scan confirmed that she had had a stroke caused by a clot in a blood vessel. The stroke coordinator gave them some written information and explained that Gladys's speech, swallowing and limb function could be expected to improve and that the physiotherapist, occupational therapist and speech and language therapist, all of whom specialized in rehabilitating people after stroke, would be involved in her care. Gladys was prescribed aspirin to reduce the possibility of further clot formation, and as her blood pressure continued to remain high, she was started on medication to lower it. She was further found to have a raised cholesterol level, for which medication was prescribed separately. Her blood glucose was also monitored while in hospital.

Gladys stayed on the medical ward for 12 days during which she made steady progress. After 48 hours, her swallowing reflex was returning and she started a diet of liquidized food. Her intravenous infusion was discontinued as she was drinking well. The physiotherapists carried out assessments of her balance and coordination and she started to move about. She was having daily speech therapy and her speech was improving. Gladys was, after the initial shock of having a stroke, very positive and determined in her outlook, and her family were extremely supportive. She was transferred to a rehabilitation unit where she stayed for a further four weeks. Her good progress continued, and after a multidisciplinary case conference and a home assessment, Gladys was discharged home. She was now walking with a stick and eating and drinking normally. She initially returned to the day hospital twice a week, where her blood pressure continued to be monitored, and she had physiotherapy and speech therapy. The community stroke nurse visited her at home once a fortnight for several months after the stroke and monitored her blood pressure, cholesterol and diabetes. Eventually, Gladys regained most of her former function and now continues to lead an independent life surrounded by family and carers – she still goes to church by herself every Sunday morning.

Poor blood circulation

Veins are equipped with one-way semilunar valves to assist the return journey of blood to the heart. However, sometimes the valves become ineffective, resulting in a partial backflow of blood, which can lead to a weakening of the vein walls. Superficial veins, predominantly in the legs, may become so elongated and dilated (varicose veins) that they are readily visible at the surface of the skin (Book 2, Section 4.2).

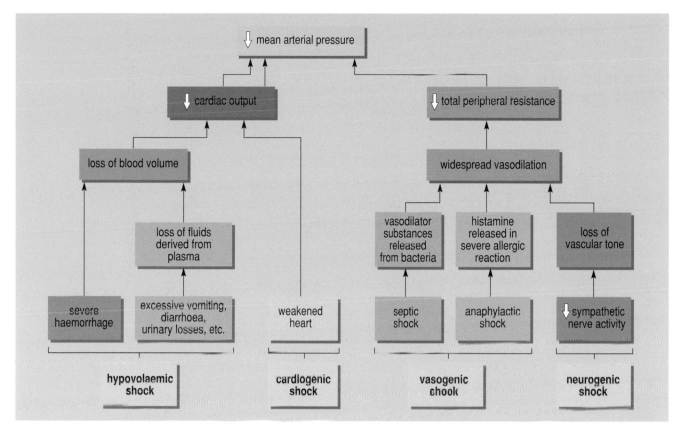

Figure 2.17 The four main types of circulatory shock.

Deep vein thrombosis (DVT) is a very serious condition where blood clots develop in the deep veins of the legs as a result of poor circulation. This may be the result of CHF (see above) or immobilization of the lower limbs for extended periods, as occurs during some forms of long distance travel. Usually DVTs are accompanied by some pain, tenderness and swelling at the back of the knee or calf. Indeed, recent newspaper articles have highlighted the fatal consequences of DVT that can occur during long-haul aeroplane flights, where passengers often remain immobile in a seated position for 5 hours or more. The cause of death in DVT is usually the result of blood clots travelling from the legs to the lungs where they cause pulmonary thromboemboli. A Department of Health statistic states that about 1% of people who develop DVT die as a direct result.

Circulatory shock

When blood pressure falls (hypotension) to a level whereby blood flow through tissues and organs is inadequate, then circulatory shock can occur. There are four main types of circulatory shock: hypovolaemic, cardiogenic, vasogenic and neurogenic (Figure 2.17).

Hypovolaemic shock is caused by a sudden loss of blood, which produces a fall in blood volume. This may result from severe haemorrhage following injury to a blood vessel (e.g. a rupture of a blood vessel after a fall downstairs). Alternatively, blood volume may be greatly reduced through the loss of plasma fluids (e.g. excessive diarrhoea or as a result of skin burns). **Cardiogenic shock** occurs when the heart is unable to pump enough blood through the circulation to maintain blood pressure. The result is that both cardiac output and arterial pressure are reduced, leading to poor tissue perfusion and possible organ failure.

Vasogenic shock (of which there are two types, septic and anaphylactic) is caused by widespread vasodilation due to the presence of vasodilator substances in the blood (Figure 2.17). Lastly, there is **neurogenic shock** which occurs when there is an abnormal reduction in sympathetic nerve activity. This can occur in response to anaesthesia, brain damage or extreme pain or stress. There is a loss of sympathetic vascular tone; this leads to peripheral vasodilation. In both vasogenic and neurogenic shock, there is a marked reduction in peripheral resistance, with a consequent reduction in blood pressure, leading to hypotension. Here again, organ failure is a possibility.

A common situation in which hypotension occurs momentarily is **orthostatic** or **postural hypotension** (Book 2, Section 4.2). Postural hypotension is a temporary condition where low blood pressure occurs due to insufficient compensatory responses to the gravitational shifts in blood distribution when a person suddenly moves from lying down to a sitting or standing position. This may happen as an older person stands up after a prolonged period in bed or from a reclined armchair.

When an individual moves from a horizontal to a vertical position, gravity causes blood to pool in the venous capacitance vessels of the legs. There is a subsequent decreased venous return (pre-load), a decreased stroke volume, a lowered cardiac output (after-load) and a lowered blood pressure. However, due to reduced use of the compensatory reflex mechanisms described in Section 2.4.6, the blood pressure may not be restored to its normal level. The resultant postural hypotension and decreased blood flow to the brain can result in loss of coordination, dizziness and even fainting (called *syncope*), predisposing the person to falls and injury (see Case Report 2.3).

Further information about cardiovascular health issues can be found at the British Heart Foundation and World Health Organization websites (see Further Reading).

Case Report 2.3 Postural hypotension and hypovolaemic shock

Mabel is an 84-year-old widow who lives alone in a three-bedroomed terraced house. Home carers visit twice daily to assist her. She has one son who lives some distance away. Mabel has epilepsy, for which she takes medication. Mabel also has a history of dizzy spells and falls and has been diagnosed as having anaemia and postural hypotension, which can be quite severe when she gets out of bed.

One morning, a carer arrived to find Mabel lying at the bottom of the stairs 'in a pool of blood'. Although conscious, Mabel had been unable to get-up off the floor. The bleeding appeared to be coming from a very large wound on the front of her scalp. The carer called an emergency ambulance. The paramedics could find no other injuries and applied a dressing with a pressure bandage to her head before taking her to the nearest A&E. As a precautionary measure, they also immobilized her neck in case she had a cervical spine injury.

In A&E, Mabel was examined thoroughly. Despite the apparent severe loss of blood, her blood pressure and pulse were just within normal limits for her age – blood pressure 145/60 mmHg and pulse 68 beats per minute. However, Mabel looked very pale. Blood samples were taken for analysis of urea and electrolyte levels and for a total blood cell count. Her blood group was identified and appropriately cross-matched blood made available for possible transfusion. An intravenous infusion of dextrose solution was started. Mabel's mental status was monitored frequently, because staff were concerned that since she could not remember the accident, she may have been unconscious for some time. A computerized tomographic (CT) scan of her head showed no bleeding into the brain and X-rays indicated that her cervical spine had not been fractured, although degenerative changes of the bones were evident. Mabel's scalp was sutured under local anaesthetic, and she was transferred to a ward for further monitoring and care.

Shortly after admission to the ward, Mabel's blood pressure suddenly dropped to 100/50 mmHg, and her pulse rose to 96 beats per minute. She appeared restless as well as looking pale and feeling clammy and cold. A blood transfusion was started and over the next 12 hours, three units of cross-matched blood were transfused into her. Mabel's condition was monitored carefully throughout this period, during which her blood pressure gradually increased and eventually stabilized. (Note that blood transfusions are preferably avoided due to the potential risks of blood-borne disease – even though all blood is screened.)

Over the next few weeks, Mabel gradually recovered from her accident and was able to be discharged back home. She continues to live alone but has increased help and support from her son, carers and close neighbours.

Summary of Section 2.6

1   In westernized societies, more people die from cardiovascular-related incidents and diseases than from any other cause.

2   Hypertension is multifactorial in origin but produces adverse structural and functional demands on the heart and blood vessels that are associated with decreased life expectancy. Serious complications of hypertension include damage to the retina, kidney failure, cerebral haemorrhage, thrombosis, aneurysm and congestive heart failure. Hypertension is commonly treated with β1-adrenoceptor antagonists (beta-blockers), thiazide and loop diuretics, ACE inhibitors and vasodilator drugs.

3   Coronary heart disease (CHD) is caused by atherosclerosis, a gradual narrowing of the arteries due to plaque formation. Consequently, the delivery of O_2 to cardiac muscle is reduced. A variety of symptoms may result, ranging from chest pain (angina) to heart attack (myocardial infarction). Coronary artery bypass surgery is sometimes performed to treat advanced CHD.

4   In congestive heart failure (CHF), cardiac output is insufficient to perfuse the body. As a result, both 'pre-load' and 'after-load' increase; to counteract this, drugs that reduce circulating blood volume (diuretics) and decrease peripheral resistance (vasodilators) are commonly prescribed. In severe cases, inotropic drugs (e.g. digoxin) are also used to increase cardiac output by increasing the force of heart muscle contraction.

5   Abnormal heart rhythms are termed cardiac arrhythmias. They are produced when the normal coordinated depolarization of heart muscle cells in the atria and ventricles is disrupted. Cardiac arrhythmias such as atrial and ventricular fibrillations or 'heart block' have characteristic patterns of ECG activity which result from different underlying causes.

6   The flow of blood through the vasculature can be affected by abnormal arterial and venous swellings (aneurysms), partial or complete blockages (thromboembolism as in DVT), poor blood circulation and vessel damage resulting in blood loss (haemorrhage).

7 Haemostasis (the process that stops bleeding) involves platelet aggregation and specific blood-borne factors (chemicals) involved in a chain of reactions ('clotting cascade') that ultimately converts fibrinogen into a stabilized fibrin net which captures erythrocytes to produce blood clots.

8 Clot formation can be influenced by anticoagulants that inhibit the conversion of fibrinogen into fibrin, antiplatelet drugs that prevent platetlet aggregation and fibrinolytic drugs (thrombolytics) that dissolve clots. These drugs are of significant value in the treatment of medical conditions where the clotting of blood may be life-threatening. An essential blood-clotting factor is lacking in haemophiliacs.

9 A fall in mean arterial blood pressure (hypotension) may occur as a result of circulatory shock. There are four main types of circulatory shock: hypovolaemic, cardiogenic, vasogenic and neurogenic.

2.7 Conclusion

Having completed this chapter, we hope that you now regard the cardiovascular system as a vital homeostatic system capable of responding and adapting to diverse biological and physical events, both in health and disease. However, it doesn't operate in isolation – its function is intimately associated and integrated, not only with the respiratory system, but also with all other organs, tissues and bodily systems.

The aim has been to give you a solid grounding in the basic anatomical components of the cardiovascular system and a functional understanding of how neural activity regulates the activity of the heart and circulation. Of major importance is a knowledge of the fundamental physiological principles (and their relationships) of how the heart pumps blood, the role of electrical activity in the cardiac cycle and the factors that affect blood flow and its distribution throughout the vasculature.

A feature strongly emphasized in this chapter is how the cardiovascular system can be affected by disease and abnormal conditions to such an extent that, as illustrated by the case reports, its function may be severely compromised. With the current awakened public attitude towards healthy living, there is no better time to adhere to the general adage: 'Look after your heart and it will look after you!'

In the next chapter we turn to the system that is the prime mover in establishing the *pneuma zotikon* ('life spirit') inside the body – the respiratory system.

Questions for Chapter 2

Question 2.1 (LOs 2.1 to 2.6)

Provide short definitions for each of the following terms: (a) haematocrit ratio, (b) vagal restraint, (c) myocardial ischaemia, (d) blood–brain barrier, (e) drug antagonist, (f) inotropic cardiac drug, (g) thromboembolism, (h) orthostatic hypotension.

Question 2.2 (LOs 2.1 to 2.3)

(a) Imagine you are an erythrocyte. Describe the pathway you flow along through the heart, naming the chambers and valves. (b) Give a short account of how electrical activity causes the heart to pump blood.

Question 2.3 (LO 2.4)

Describe and contrast the electrocardiogram (ECG) of a normal adult person with that obtained from a patient with 3 : 1 AV heart block.

Question 2.4 (LO 2.5)

Assume that your heart rate is 70 beats per minute and your stroke volume is 70 ml. (a) Calculate your cardiac output (in litres per hour). (b) What factors affect cardiac output?

Question 2.5 (LO 2.5)

Suppose you are a GP and a patient in your surgery has a sustained blood pressure of 165 / 100 . Explain what these measurements indicate and detail what types of drugs you could prescribe to alter your patient's blood pressure.

References and Further Reading

References

Allen, K., Blascovich, J. and Mendes, W. B. (2003) Cardiovascular reactivity and the presence of pets, friends, and spouses: the truth about cats and dogs, *Psychosomatic Medicine*, **64** (5), 727–739.

Dickens, C.M., McGowan, L., Percival, C., Douglas, J., Tomenson, B., Cotter, Heagerty, L.A. and Creed, F.H. (2004) Lack of a close confidant, but not depression, predicts further cardiac events after myocardial infarction, *Heart*, **90** (5), 518–522.

Mills, B. D. (1989) 80,000 pills: A personal history of hypertension, *British Medical Journal*, **298**, 445–448.

Nissen, S., Tsunoda, T., Tuzcu, E. T., Schoenhagen, P., Cooper, C. J., Yasin, M., Eaton, G. M., Lauer, M. A., Sheldon, W. S., Grines, C., Halpern, S., Crowe, T., Blankenship, J. C. and Kerensky, R. (2003) Effect of recombinant ApoA-1 Milano on coronary atherosclerosis in patients with acute coronary syndromes, *Journal of the American Medical Association*, **290** (17), 2292–2294.

Further Reading

Aaronson, P. I. and Ward, J. P. T. (2003) *The Cardiovascular System at a Glance*. Oxford: Blackwell Sciences Ltd.

British Heart Foundation, 14 Fitzhardinge Street, London, W1H 4DH, Discussion topics [online]: http://www.bhf.org.uk/questions [*Select* 'Medical' *category*] (Accessed November 2004)

British National Formulary [online]: http://www.bnf.org/bnf/bnf/current/openat/index.htm (Accessed November 2004)

Cardiopulmonary Resuscitation (CPR) – The Harvard Medical School Family Health Guide [online]: http://www.health.harvard.edu/fhg/firstaid/CPR.shtml (Accessed November 2004)

Cardiovascular Diseases – World Health Organization [online]: http://www.who.int/cardiovascular_diseases/en/ (Accessed November 2004)

Lye, M. and Donnellan, C. (2000) Heart disease in the elderly, *Heart*, **84**, 560–566.

Rang, H. P., Dale, M. M. and Ritter, J. M. (1996) *Pharmacology* (3rd edn). London: Churchill Livingstone.

Remmea, W. J. and Swedbergb, K. (2002). Comprehensive guidelines for the diagnosis and treatment of chronic heart failure, *European Journal of Heart Failure*, **4** (1), 11–22.

CHAPTER 3

RESPIRATION

Learning Outcomes

After completing this chapter, you should be able to:

3.1 Describe and illustrate the main anatomical features of the respiratory system.

3.2 Outline the cellular mechanisms of O_2 and CO_2 transport in the blood.

3.3 Illustrate specific lung diseases and their treatments.

3.4 Describe the integrative control of both the respiratory and circulatory systems.

3.5 Give examples of cardiorespiratory adaptation to altered internal and external environments.

3.1 Introduction – from oxygen to energy

The vital bellows...

In Chapter 2, we learned in detail about the activity of the cardiovascular system in both health and disease. Here we explore further the respiratory system, whose structure and function are closely integrated – through homeostatic mechanisms – with those of the cardiovascular system. As we shall discover, the coordinated functioning of these two systems is essential to life. Indeed, in his *Lectures on the Whole of Anatomy* (1653), William Harvey wrote that, with regard to the 'pre-eminence [of the lung], nothing is especially so necessary...Life and respiration are complementary. There is nothing living which does not breathe, nor anything which breathing, which does not live'.

On average, we breathe over 8000 litres of air per day. Oxygen (O_2) in this air enters the body via the respiratory system and is taken up by the cardiovascular system for subsequent delivery to all the tissues in the body; carbon dioxide (CO_2) passes from these tissues into the cardiovascular system and is then transported to the external environment via the respiratory system.

The exchange of gases between the external environment and the body is termed respiration. More precisely, respiration can be divided into two processes: external respiration and cellular respiration. **External respiration** is the exchange of O_2 and CO_2 from the body as a whole. In contrast **cellular respiration** is the chemical process in cells whereby six molecules of oxygen ($6O_2$) combine with a molecule of glucose ($C_6H_{12}O_6$) to release energy in the form of the high energy molecule adenosine triphosphate (ATP). In this chemical reaction six molecules of both CO_2 and H_2O are formed as by-products.

$$6O_2 + C_6H_{12}O_6 = 6CO_2 + 6H_2O + ATP \text{ (energy)}$$

Since living cells require a continuous source of O_2 for metabolism and need to expel the CO_2 produced in the process, there needs to be a continuous exchange of gases with the environment. Breathing (or ventilation) is the mechanical activity of moving air into and out of the lungs; it has two phases, *inspiration* and *expiration*. Before we consider the function of the respiratory system we will look at its anatomy.

3.2 The respiratory system

The respiratory system consists of:

- the lungs (the gas-exchange organ);
- a system of airways which deliver the air to the blood vessels of the lungs;
- the musculo-skeletal system which ventilates the lungs.

3.2.1 Structure of the respiratory system

Figure 3.1 shows the main structures of the respiratory system. Air enters the body through the nostrils and passes into the nasal cavities. Normally, one nostril is slightly larger than the other, producing asymmetrical air-flow into the nose. Although perhaps not immediately obvious, nostril size alternates throughout the day in what is termed the 'nasal cycle' (Qian et al., 2001). The lining of the nasal cavity is formed from epithelial cells which are covered with cilia (fine hair-like processes) and mucus-producing cells. The mucus (or phlegm) is sticky and so traps dirt and other particles that are inhaled with the air; the cilia form a continuously beating 'conveyor belt' which moves the mucus to the throat where it is swallowed with saliva and passes into the digestive system. This process protects the rest of the respiratory system from invading particles. The air that passes through the nasal cavities is warmed to body temperature and is moistened by the mucus. From the nasal cavities, the air passes to the **pharynx** at the back of the mouth where it is joined by air that has entered the system through the mouth.

● What is the advantage of breathing through the nose rather than the mouth?

● Air that enters through the nose is filtered by the cilia and mucus. The air entering the mouth is not filtered, so any particles in the air entering the body by this route will pass unobstructed into the respiratory system. Also, breathing through the nose allows the inhaled air a greater opportunity to reach body temperature.

● What other function is served by the passage of air through the nose?

● Air in the nose can be analysed by the olfactory receptors to detect any air-borne chemical signalling molecules (**pheromones**) thereby providing input to our olfactory system.

At the base of the pharynx (throat) are two openings; one leads to the **larynx** (or 'voice-box') and the rest of the respiratory system, and the other is the entrance of the oesophagus, near the beginning of the digestive tract. Obviously it is essential that these two openings are kept separate; otherwise, when we eat or drink, food particles or fluids would enter the respiratory system and cause choking. To prevent this, there is a small flap of tissue, the **epiglottis**, which closes the larynx during swallowing, preventing food or water from passing into

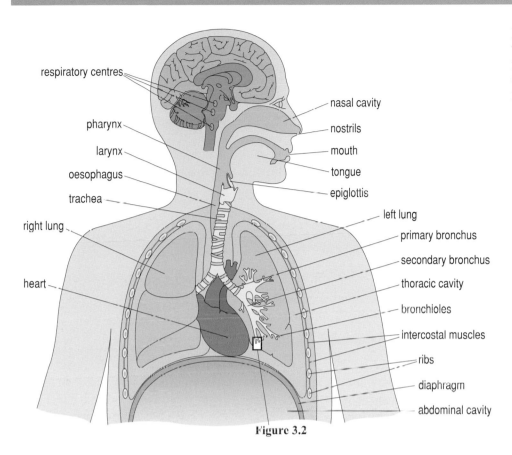

Figure 3.1 The components of the respiratory system. The microstructure of the lung and associated vasculature are shown in greater detail in Figure 3.2.

Figure 3.2

the respiratory system (Book 1, Figure 4.9). If this 'diversion system' fails and unwanted particles enter the larynx, then a cough reflex is initiated, expelling the unwanted particles.

● What happens if the cough reflex fails to remove a large respiratory blockage?

● If the cough reflex does not remove a large blockage of the respiratory tract, then choking will follow and eventually death by asphyxiation will occur. If the object is small, e.g. a peanut, it may be sucked further into one of the branches of the respiratory system and obstruct the area of lung tissue beyond that point, thereby reducing the surface area able to engage in respiratory exchange.

The larynx is also called the 'Adam's apple' and contains the vocal cords. These are epithelial folds which vibrate and produce sound as air passes over them. *Laryngitis* is an inflammation of the vocal cords caused by excessive overuse or infection. From the larynx, air passes into the **trachea** or windpipe, a hollow tube kept permanently open by rings of cartilage. The trachea divides into two branches called **bronchi** (singular, *bronchus*). These serve the left and right lungs. Like the trachea, the walls of the bronchi contain cartilage which prevents their collapse. The respiratory tract (similar to the way in which the circulatory system proceeds from a few wide-bore main vessels into a network of numerous narrow-bore capillaries) divides into increasingly fine tubes – similar to the 'trunk → branch → twig' progression of a large tree. Each main bronchus divides into two smaller tubes, which each divide into two yet smaller tubes and so on, a further 14 times, forming increasingly smaller bronchi, then **bronchioles**, and finally terminal bronchioles – there are over 1500 miles of airways in the average human lung! The bronchi, primary bronchioles and terminal bronchioles conduct gas to and from the exterior and thus form the **conducting zone**. The terminal bronchioles each divide a further seven times into respiratory bronchioles, then **alveolar ducts** and finally into alveolar sacs (alveoli; singular, alveolus) – which represent the ends of the 'bronchial tree'. The normal lung contains an average of over 300 million alveoli. Gas exchange occurs in the respiratory bronchioles, alveolar ducts and the alveoli, which thus form the **respiratory zones**. This repeated branching serves to greatly increase the total area over which gas exchange can occur and effectively results in a thin film of blood of surface area about 400 square metres (roughly equivalent to the size of a large tennis court) in contact with the alveolar air and potentially available for gas exchange into and out of the blood.

The alveoli are surrounded by pulmonary capillaries. Oxygen from within the alveoli diffuses into the pulmonary capillaries, while CO_2 diffuses from the blood into the alveoli. Since the walls of an alveolus and a pulmonary capillary are each only one cell thick, the gases have only to diffuse through a two-cell-thick layer (Figure 3.2) and this facilitates rapid diffusion and highly effective gas exchange.

The efficiency of gas exchange in the lungs is further increased by the thin liquid film that lines the surface of each alveolus. However, at any air-water interface (or surface), water molecules are more strongly attracted to neighbouring water molecules than to air molecules. (Think of drops of water at the end of the kitchen tap). Consequently, this unequal attraction creates a force known as **surface tension** which tries to keep water molecules together and opposes any forces that try to separate them – which is a biological engineering problem for a structure, like an alveolus, that is designed to expand in size when inflated with air. A clue to the answer provided by nature lies in the image of a child blowing bubbles using slightly soapy water. The surface tension of a thin liquid film can be reduced by adding a very small quantity of detergent which reduces the surface tension and allows the water to be easily expanded into the transparent spherical shape of a bubble.

In the lungs, nature's answer is for the walls of the alveoli to be composed of two different types of epithelial cell. The first type are similar to the endothelial cells of the capillaries and the second produce a '*surf*ace *ac*tive age*nt*' (*surfactant*), which is a detergent-like fluid that adds to the thin liquid film lining the alveolar walls. Surfactant reduces the surface tension inside the alveoli thereby allowing the lungs to expand when inflated. Surfactant also serves to prevent the collapse of the alveoli and is particularly important in allowing the

Figure 3.2 Schematic diagram of alveoli in contact with pulmonary capillaries.

lungs of newborn babies to inflate when they take their first breaths. Premature babies born before their surfactant production system is fully functional suffer from **respiratory distress syndrome** (RDS). Surface tension in the lungs of these babies is high and many alveoli fail to expand. Failure to produce enough surfactant may also be a problem in adult life; for example, surfactant production in the lungs of smokers is greatly reduced, increasing the likelihood of breathing difficulties compared to non-smokers (see Section 3.3.5).

There are no cilia or mucus-producing cells lining the walls of the respiratory zone, so any particles that may have escaped the filtering system higher up in the lungs will be deposited in the alveoli. This occurs in individuals routinely inhaling small particulate matter, e.g. smokers, people exposed to asbestos fibres and miners exposed to coal dust. All these particles are so small (less than 10 μm, i.e. one-hundredth of a millimetre) that they are not trapped in the mucus of the upper respiratory tract. Some of these particles may be phagocytosed (engulfed) by white cells (see Book 1, Figure 2.23) in the lungs and transported to the lymph nodes of the lungs where they accumulate as black deposits. Those particles that remain in the respiratory zone continue to cause irritation to the lungs and may eventually cause chronic disease.

3.2.2 Respiratory mechanics – 'Every breath you take...'

The lungs are located within the thoracic (chest) cavity. This space is formed by the chest wall of the ribcage, the *sternum* (breastbone), and the muscular *diaphragm*, which separates the thoracic cavity from the abdominal cavity. Both the lungs and the chest wall are elastic structures which slide against one another on expansion and deflation. This action is made possible by the **pleura**, thin membranes that cover the lungs and the inside of the chest wall. A thin layer of fluid fills the space between the lung pleura and the chest wall pleura (known as the *pleural cavity*). The lubricated membranes and the fluid layer allow the lung to move easily within the thoracic cavity, and prevents the collapse of the lung by maintaining the surface tension that holds it to the chest wall. Infections of the pleural cavity cause inflammation of the pleural membranes, a disease called **pleurisy**.

Inspiration is an active process which involves the diaphragm and the external and internal muscles between the ribs, the **intercostal muscles**. In its resting state, the diaphragm is dome-shaped, arching upwards into the thoracic cavity. On contraction, the diaphragm flattens out and moves downwards, increasing the air intake capacity by up to 75% (see Figure 3.3a, overleaf). The vertical distance moved by the diaphragm is 1.5 cm during shallow breathing to as much as 7cm during deep inspiration. The majority of the rib movement during breathing is achieved by the external intercostal muscles. On inspiration, the ribs, which are attached to the vertebral column and the sternum, are moved upwards and outwards by contraction of the external intercostal muscles. Contraction of the diaphragm and the external intercostal muscles increases the volume within the ribcage, so there is a large area for the lungs to expand into. The movement of the diaphragm and ribcage during inspiration and expiration are shown diagrammatically in Figure 3.3. Because the pleural membranes covering the lungs are so tightly stuck to the chest wall by the film of fluid (due to surface tension), expansion of the chest wall will cause expansion of the lungs. Therefore, as the chest expands, the air pressure within the lungs decreases, air flows in through the trachea and thus forces the alveoli to expand and fill with air.

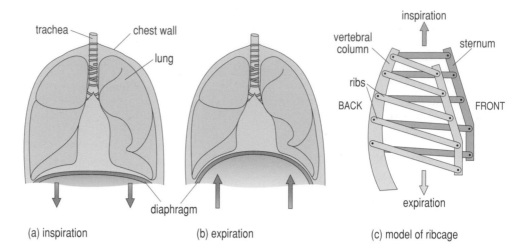

(a) inspiration (b) expiration (c) model of ribcage

Figure 3.3   The mechanics of inspiration and expiration. (a) The diaphragm flattens and moves downwards, so increasing the thoracic capacity and allowing air to enter the lungs. (b) During expiration, the diaphragm curves upwards and the thoracic capacity is reduced. (c) A model of the ribcage in side view. During inspiration, the ribs and sternum move upwards and outwards; during expiration, the ribs and sternum move downwards and inwards.

●   What do you think would happen if this film of fluid was disrupted – say, if the lung was punctured, allowing an air bubble to form between the lung and the chest wall?

●   If there was an air bubble between the lung and the chest wall, it would break the surface tension holding the lung to the chest wall and make lung inflation, and therefore inspiration, difficult.

How do you breathe? Place your hands over your ribcage and try to feel the difference between the muscles contracting when you breathe normally and those that contract as you force out a breath, for example, when trying to blow up a balloon.

Expiration is generally a passive event brought about by relaxation of the diaphragm and external intercostal muscles. The ribcage and diaphragm return, by natural recoil, to their original pre-inspiratory positions. The consequent retraction of the chest wall forces air out of the lungs. Active, or forced, expiration (see above) is mainly achieved by contraction of the internal intercostal muscles, aided to some extent by contraction of the abdominal muscles.

The volume of air breathed in and out with every breath is known as the **tidal volume**. It is normally around 0.5 litres. If you take a deep breath in, i.e. inspire as hard as you can, the extra volume of air inspired is the *inspiratory reserve volume*. Likewise, if you breathe out as hard as you can after a normal intake of breath, the extra volume breathed out is the *expiratory reserve volume*. There is always a small amount of air left in the lungs in addition to the expiratory reserve volume and this is known as the *residual volume*. Average values of lung volumes for men and women are given in Table 3.1.

Table 3.1 Average values of lung volumes in litres. Note the large differences between the inspiratory reserve volume and total lung capacity values for men and women. Tidal volume is the same in men and women but lung capacity is much greater in men due to their generally larger chest size.

	Men	Women
tidal volume (litres)	0.5	0.5
inspiratory reserve volume (litres)	3.3	1.9
expiratory reserve volume (litres)	1.0	0.7
residual volume (litres)	1.2	1.1
total lung capacity (litres)	6.0	4.2

● What single factor do you think will provide a sustained improvement in respiratory function?

● Regular aerobic exercise (see Section 3.5.1 below).

An important index of lung function is the **peak flow rate**, also known as the peak expiratory flow (PEF) rate. After an inspiration to the normal total lung volume, a person is asked to breathe out as fast as possible (forced expiration) through a peak flow meter. Peak flow rate measures two parameters: firstly, the muscular ability to expel air from the lungs; and secondly, how well the air is able to pass through the conducting air passages. The maximum flow rate is normally reached in the first tenth of a second of the forced expiration and the flow rate is subsequently calculated in litres per minute. Healthy young adults (18–25 years) achieve flow rates of about 480–580 litres per minute; but this value varies greatly with age, gender, physical stature and of course health. Any obstructions to the passage of air through the lungs, for example by abnormal constriction of the airways due to illness, will greatly affect peak flow rate readings. Obstruction to air flow is commonly associated with asthma, chronic bronchitis or emphysema (see Section 3.9.1).

3.2.3 Gas exchange in the lungs and tissues

Oxygen from inspired air that has reached the alveoli diffuses into the blood of the capillaries that surround the alveoli. Carbon dioxide in the bloodstream diffuses into the alveoli and is removed on expiration (Figure 3.4). The concentration of a particular gas in a mixture of gases can be expressed as the partial pressure of that gas. The atmospheric pressure, i.e. the pressure exerted by the gases in the Earth's atmosphere has a value of 760 mmHg (see Box 2.1 in Section 2.4.8). Oxygen accounts for about 21% of the Earth's atmosphere so the partial pressure of O_2 (P_{O_2}) in the atmosphere is 0.21×760 mmHg = 160 mmHg. Although the P_{O_2} of the inspired air is 160 mmHg, the P_{O_2} in the alveoli is lower (100 mmHg).

● What does the lower partial pressure of O_2 in the alveoli mean?

● It means that there is a lower percentage of O_2 in the alveolar air.

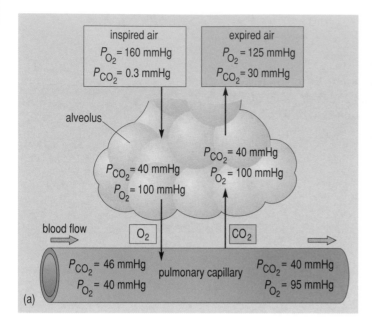

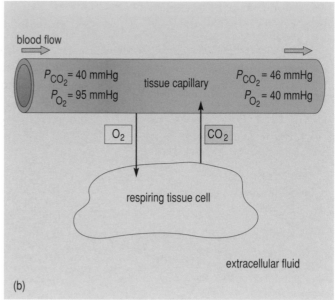

Figure 3.4 Gas exchange at (a) the lungs and (b) the tissues, showing the partial pressures of the gases, P_{O_2} and P_{CO_2}. (*Note*: in reality cells lie adjacent to blood vessels and are more densely packed than shown here.)

● Can you suggest why the P_{O_2} in the alveoli is less than the P_{O_2} of the inspired air?

○ Not all of the inspired air reaches the alveoli because of the volume of air that makes up the residual volume. The residual air mixes with the inspired air, reducing the P_{O_2} of the air in the alveoli and thus ensuring that the P_{O_2} in the alveoli is always less than that of the air entering the lungs.

The P_{O_2} in deoxygenated blood is much lower than in the alveoli, because O_2 has been removed from the blood by the tissues. The P_{O_2} in blood entering the lungs is 40 mmHg; thus a partial pressure (i.e. concentration) gradient exists between alveolar gas and the blood, so O_2 diffuses into the blood, giving a P_{O_2} for the blood leaving the lungs of 95 mmHg. There is also a small partial pressure gradient for CO_2 between the blood and the alveolar gas. The P_{CO_2} of deoxygenated blood is 46 mmHg, whereas that of the alveolar gas is 40 mmHg. The P_{CO_2} of the blood leaving the lungs is 40 mmHg, indicating that equilibrium is reached between the blood and alveolar gas, and demonstrating the ease with which CO_2 moves across membranes.

● Is equilibrium reached between the O_2 in the alveoli and that in the blood?

○ No. The P_{O_2} in the alveoli is 100 mmHg whereas in the blood leaving the lungs it is 95 mmHg. This suggests that O_2 is not as soluble in the blood as CO_2.

The inability of O_2 to reach equilibrium is due to its method of transport in the blood, which is described in the next section.

3.2.4 Oxygen transport in the blood

Although O_2 diffuses into the blood, it is not very soluble in plasma; so how does the blood accommodate such a high concentration of O_2 without it coming out of solution and forming bubbles? The ability of blood to carry O_2 is due to the presence of the respiratory pigment, haemoglobin. There are several types of haemoglobin. In adults, haemoglobin (Hb for short) is a protein formed from four polypeptide chains, called *globins* – there are two alpha and two beta chains.

Attached to the centre of each globin chain is a small non-protein structure known as a *haem group*. The haem group has at its centre an iron ion (Fe^{2+}) which will pick up one O_2 molecule. Since there are four globin chains and four haem groups each with one Fe^{2+} (Figure 3.5), one haemoglobin molecule can carry four O_2 molecules. (A three-dimensional model of human haemoglobin, but without the haem groups, is shown in Figure 2.5 of Book 1.)

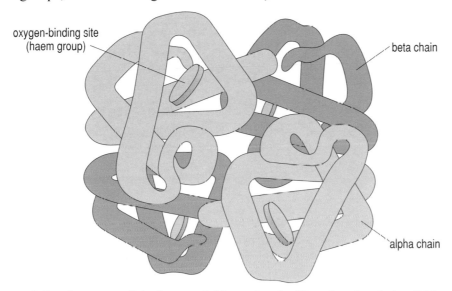

Figure 3.5   Structure of the haemoglobin molecule. Note that the chains fold up to form globular structures – this is why they are known as globins.

When O_2 is bound to haemoglobin, the haemoglobin is said to be oxygenated and the complex formed is called **oxyhaemoglobin**. The binding of O_2 to all four globin chains of a haemoglobin molecule can be written in terms of a simple chemical equation:

$$\underset{\text{haemoglobin}}{\text{Hb}} + 4O_2 \rightleftharpoons \underset{\text{oxyhaemoglobin}}{\text{Hb}(O_2)_4}$$

Recall that the two-way arrows indicate that this reaction can take place in both directions; in other words, it is reversible. So in a situation where there is a plentiful supply of O_2 (in the capillaries surrounding the alveoli of the lungs), the reaction will proceed from left to right and oxyhaemoglobin will be formed. Here, the blood haemoglobin is almost 98% saturated with O_2. However, where the amount of O_2 available is low (in the capillaries within the tissues), then the reaction will proceed from right to left and O_2 will be released from the oxyhaemoglobin, which reverts back to haemoglobin. This may seem straightforward: where there is abundant O_2 in the cardiovascular system it is picked up by haemoglobin, and where there is little O_2 it is released from haemoglobin.

However, as with many other biological phenomena, the binding of oxygen to haemoglobin is more complicated than this. Figure 3.6a demonstrates graphically the binding of O_2 to haemoglobin at normal physiological pH (=7.4) – the graph is

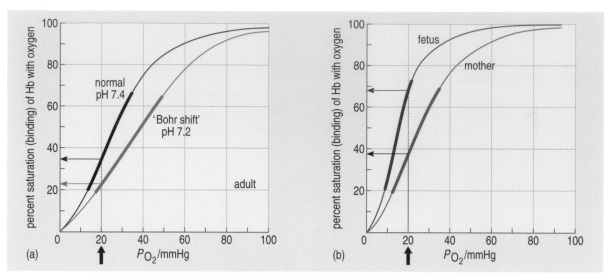

Figure 3.6 Oxygen–haemoglobin dissociation curves. The steepest parts of each curve (shown thickened) represent the most efficient regions of O_2 and haemoglobin (Hb) binding. (a) A normal adult dissociation curve (at 37 °C, pH 7.4) and the effect on the position and shape of the curve of increasing the acidity of the blood (pH 7.2). (b) Dissociation curves for a mother and fetus (at 37 °C, pH 7.4).

known as the **oxygen–haemoglobin dissociation curve**. The vertical axis indicates the amount of oxyhaemoglobin present in the blood as a percentage of the total haemoglobin. Note that the shape of the graph is not a straight line over all its range – there are places where oxygen and haemoglobin do not bind together in a 'linear' fashion. The graph is an '*S*'-shaped (or *sigmoid*) curve. The steepest part of the curve shown in Figure 3.6a represents the most efficient range over which haemoglobin and O_2 interact at a given P_{O_2} in the blood.

The association of oxygen and haemoglobin is affected by several factors – the two most important are blood pH and temperature. Variations in these factors greatly influence the efficiency of oxygen uptake by haemoglobin. For example, when the blood becomes more acidic (lower pH), there is a lateral shift in the oxygen–haemoglobin dissociation curve to the right, called the **Bohr shift**, which results in a decrease in the affinity of haemoglobin for O_2. Figure 3.6a demonstrates this shift – at a P_{O_2} of 20 mmHg (black vertical arrow), haemoglobin is about a third less well saturated with O_2 at comparatively acidic pH 7.2 compared with the normal pH 7.4.

● What effect do you think temperature has on the oxygen–haemoglobin dissociation curve?

○ At higher temperatures, haemoglobin is less saturated with oxygen.

Figure 3.6b shows the oxygen–haemoglobin dissociation curve for adults (adult haemoglobin) and for the fetus, which has a different type of haemoglobin (fetal haemoglobin). Fetal haemoglobin has two gamma chains in place of the beta chains of adult haemoglobin, endowing it with a higher affinity for O_2 than the adult form. Consequently, the oxygen–haemoglobin dissociation curve for fetal haemoglobin is shifted to the left, indicating a greater affinity for O_2 than adult haemoglobin. As shown in Figure 3.6b, for a given P_{O_2} of 20 mmHg (black arrow), fetal haemoglobin is nearly twice as saturated with O_2 as the maternal type (see also Book 4, Chapter 1 for further discussion of these haemoglobins).

● What is the advantage to the fetus of having a form of haemoglobin with a greater O_2 affinity than the mother's haemoglobin?

● The difference in O_2 affinities favours the transfer of O_2 from maternal to fetal haemoglobin. The advantage to the fetus is that its haemoglobin will readily pick up and become saturated with the O_2 transported in the mother's blood.

The capacity of the blood to carry O_2 is also greatly reduced by carbon monoxide (CO), a gas emitted by car exhausts and faulty gas appliances. Carbon monoxide competes with O_2 for the haemoglobin and in this chemical competition, carbon monoxide always 'wins' because it binds much more strongly than O_2. Carbon monoxide and haemoglobin combine, *irreversibly*, to form **carboxyhaemoglobin** (HbCO). Inhaling carbon monoxide will therefore progressively reduce the amount of haemoglobin available to bind O_2. Carbon monoxide poisoning results, whereby very little O_2 is available for cellular respiration and, if the source of CO is not removed, death could result due to the total lack of oxygen (**asphyxiation**). In mild cases of CO poisoning, recovery occurs gradually, as fresh Hb is manufactured.

3.2.5 Carbon dioxide transport in the blood

Carbon dioxide is carried in the blood in a number of forms:

- Physically dissolved in blood plasma (10%), but this depends on the difference in P_{CO_2} concentrations between the tissues and blood plasma.

- Bound to haemoglobin (30%) to form *carbaminohaemoglobin* (HbCO$_2$); however, CO_2 only binds with the globin portion of haemoglobin (in contrast to O_2).

- As bicarbonate (HCO_3^-) ions; about 60% of the CO_2 is carried dissolved in the plasma as carbonic acid (H_2CO_3) or bicarbonate (HCO_3^-), according to the equation below:

$$CO_2 + H_2O \underset{\text{carbonic anhydrase}}{\rightleftharpoons} H_2CO_3 \rightleftharpoons HCO_3^- + H^+$$

carbon dioxide   water   carbonic acid   bicarbonate ion   hydrogen ion

● What effect will increased CO_2 levels have on the pH of the blood?

● The blood pH will be reduced. High CO_2 levels will produce more carbonic acid that will dissociate into bicarbonate ions and H^+ ions. The H^+ ions will make the blood more acidic and therefore lower the pH (typically from 7.4 to 7.2).

The reaction between CO_2 and water is normally very slow; however, in erythrocytes there is an enzyme, **carbonic anhydrase**, which greatly speeds up this reaction. The concentration of CO_2 in the tissues is much higher than the concentration in the erythrocytes and so a CO_2 partial pressure (P_{CO_2}) gradient exists between tissue and erythrocyte. Figure 3.7 (overleaf) shows the reactions that take place in an erythrocyte when it reaches the tissue. Carbon dioxide diffuses into the cell and, due to the presence of carbonic anhydrase, reacts immediately with water to produce carbonic acid. This conversion reduces the CO_2 concentration inside the erythrocyte, thereby maintaining the P_{CO_2} gradient. As a result, 90% of the CO_2 in the circulation enters the erythrocytes. Blood spends on average 0.75 seconds in the capillary and so the time available to pick up a full load of CO_2 from the tissues is very short. The presence of carbonic anhydrase allows the rapid uptake of CO_2 into the blood.

Figure 3.7 Carbon dioxide transport between the tissues and erythrocytes.

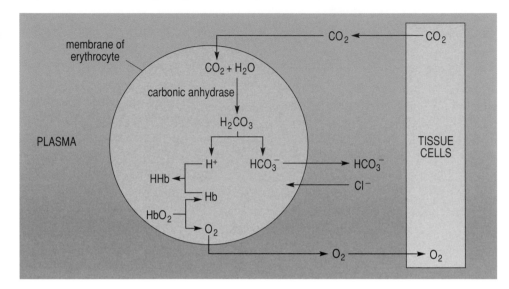

● Can you suggest another reason why it is necessary to retain CO_2 in the erythrocytes, rather than in the plasma?

● CO_2 is then readily accessible in the erythrocytes for removal in the lungs. Also, if CO_2 was allowed to accumulate in the plasma and react with water, the pH of the blood would fall, which would not be desirable, as it is important that blood pH values are maintained within strict limits. A change in pH would affect a multitude of other chemical reactions related to metabolic activities – particularly those that depend on the activity of enzymes (see Book 1, Section 2.3.2).

Within the erythrocytes, the carbonic acid (H_2CO_3) dissociates into a bicarbonate ion (HCO_3^-) and a hydrogen ion (H^+). If the H^+ ions were to accumulate inside the erythrocyte they would eventually kill it, but the ions are biochemically buffered (neutralized) by haemoglobin, thereby keeping their concentration within safe limits. The presence of H^+ ions encourages the displacement of O_2 from the oxyhaemoglobin (called the **Haldane effect**). The O_2 then diffuses out of the erythrocyte and into the tissue – especially in metabolically active tissues (e.g. exercising muscle), where the concentration of CO_2 will be high and the tissue in need of further supplies of O_2. A rather neat 'gaseous handshake' – to give O_2 with one hand and to take away CO_2 with the other. Put more scientifically, unloading O_2 from haemoglobin at oxygen-requiring tissue sites increases the haemoglobin's ability to pick up CO_2.

As already mentioned, the entry of CO_2 into the erythrocyte and its reaction with water produces HCO_3^- ions. The erythrocyte membrane is permeable to these negatively charged ions, which diffuse out into the plasma. This outward movement of negative charge results in an excess of positive charge inside the cell. But the erythrocyte membrane is not permeable to positively charged ions, so positive charge cannot leave the cell to maintain electrical neutrality. Instead, there is an inward movement of chloride (Cl^-) ions – normally abundant in the plasma – known as the **chloride shift**, which compensates for the loss of HCO_3^- ions from the erythrocyte.

3.2.6 The control of respiration

The control of respiration is brought about by innervation of the inspiratory muscles. Generally, respiration is an involuntary, automatic event. We are not aware of it happening unless we try to hold our breath, or breathe out deeply, when control becomes voluntary. Two separate areas of the brain are responsible for the involuntary (automatic) and voluntary control over respiration. An area in the pons and medulla at the base of the brain is the site of automatic (or reflexive) control which operates via the autonomic nervous system (see Figure 1.10 and Section 1.4.3 of Book 2). In addition, brain areas allowing for the voluntary control of respiration via the somatic nervous system are found in the cerebral cortex. Each of these centres receives information about the respiratory status of the body – from muscles and tissues via nerves, and from the blood by chemical stimuli – and sends out impulses to the respiratory muscles to increase or decrease the rate of breathing as appropriate.

The diaphragm is innervated by the phrenic nerve (originating in the respiratory centre in the brain's medulla), and the intercostal muscles receive inputs from the thoracic nerves. Information from the medulla produces a regular cycle of activity in these nerves, controlling the rate of ventilation. It can be said that the respiratory centres act as the 'respiratory pacemaker'. There are two types of neuron in the respiratory centres which both send impulses to the phrenic and thoracic nerves. The first set are activated (or fire) during inspiration (*the inspiratory neurons*) and the second set of neurons fire during expiration (*the expiratory neurons*). The inspiratory neurons inhibit the firing of the expiratory neurons and vice versa.

When we breathe in, the inspiratory centre in the medulla is active and the inspiratory neurons are firing to stimulate contraction of the diaphragm and the intercostal muscles. Inspiration is terminated by two physiological events: (a) activity in the respiratory centre, which inhibits breathing; and (b) information from **stretch receptors** in the lungs, which tell the brain when the lungs have expanded to full capacity. The inspiratory neurons stop firing and relieve the block on the expiratory neurons, which leads to relaxation of the inspiratory muscles. The respiratory centre therefore acts as a 'pattern generator' for respiration. The pattern generator can be modified in response to changes in the metabolic requirements of the body, which are monitored by **chemoreceptors**, receptors that respond to the chemical environment. There are two sorts of chemoreceptor involved with the control of respiration – central and peripheral chemoreceptors. Central chemoreceptors in the medulla are a group of modified nerve cells that respond to the concentration of H^+ ions in the cerebrospinal fluid. If the concentration of H^+ ions increases, these cells become more active, increasing the activity of the respiratory centre and increasing ventilation.

● Under what circumstances would the H⁺ ion concentration of the extracellular fluid increase?

● If there is an increase in the partial pressure of CO_2, the gas will pass into the plasma where it will react with water, resulting in the formation of HCO_3^- ions and H^+ ions.

Peripheral chemoreceptors are specialized groups of cells found in the same location as the baroreceptors, i.e. at the branch point of the carotid arteries and in the walls of the aorta (Section 2.4.6). These sensory cells respond to decreases in O_2 concentration of the blood and send the information to the respiratory centres in the CNS to stimulate ventilation. Their activity will be discussed in greater detail in the next section. Figure 3.8 is a schematic diagram of the physiological control of ventilation.

Figure 3.8 The interrelationships between neural, chemical and pulmonary information controlling ventilation.

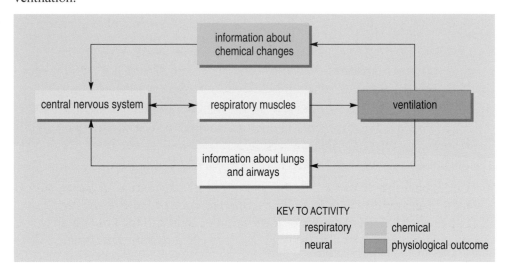

3.2.7 Non-respiratory functions

The respiratory system also performs important non-respiratory functions, for example:

• By acting as a simple bellows it allows speech, singing and other vocalizations to be made. The two bands of elastic tissue that lie across the opening of the larynx, called the **vocal cords**, can be stretched and positioned into different shapes by the laryngeal muscles. As air is passed over the vocal folds they can be made to vibrate to produce characteristic patterns of sound.

• Air drawn into the nose from the external environment can be sampled by olfactory (smell) receptor cells in the olfactory nasal mucosa to detect any airborne chemical messengers (pheromones).

• The respiratory tract provides a means for water loss and heat elimination. (After strenuous walkies, watch your family pet dog pant deeply as part of its thermoregulation.) As previously mentioned, inspired atmospheric air can be humidified and warmed by the respiratory airways; this is essential to prevent the alveolar membranes from drying out, for O_2 and CO_2 are unable to diffuse through dry membranes.

- During inspiration, there is a fall in pressure in the chest cavity, which reduces the resistance of blood vessels, thereby facilitating the flow of blood around the body. In a similar way, respiratory movements also aid the movement of lymph through the lymphatic system.

- The respiratory system provides defence mechanisms against inhaled particulate matter (e.g. dust, bacteria); these mechanisms include nasal hair and cilia lining the airways (see Section 3.2.1). Further respiratory defence mechanisms include coughing and sneezing. Moreover, an army of phagocytic scavenger cells (alveolar white cells) 'roam' the air sacs in the lungs.

Summary of Section 3.2

1   The respiratory system allows the passage of O_2 from the external environment to the internal environment. It also allows the release into the external environment of CO_2, a potentially toxic by-product of respiration.

2   The respiratory system consists of the lungs, a branching series of tubular airways, and the ribcage, intercostal muscles and diaphragm whose movements allow the filling and emptying of the lungs.

3   Oxygen diffuses from inspired air into the capillaries and it is carried in the blood bound to the respiratory pigment haemoglobin. The oxygen–haemoglobin dissociation curve predicts the uptake and release of O_2 and is significantly altered by several factors, e.g. pH and temperature.

4   Carbon dioxide diffuses from the tissues into the capillaries. In the erythrocyte, CO_2 reacts rapidly with water due to the presence of the enzyme carbonic anhydrase. The CO_2 and water form carbonic acid and then bicarbonate and the majority of the CO_2 in the blood is carried in this converted form.

5   Respiration is controlled by the respiratory centre in the brain, which modulates the activity of inspiratory and expiratory neurons, which effect inspiration and expiration, respectively.

6   The respiratory centre receives information from stretch receptors in the lungs, which inform the brain when the lungs are fully expanded, and also from peripheral and central chemoreceptors, which monitor the pH and O_2 content of the blood.

7   The respiratory system performs additional functions such as vocalization, smell, the facilitation of blood and lymph flow, thermoregulation, and defence against inhaled particles and organisms.

3.3 Diseases that can affect the lungs

Diseases affecting the respiratory system are responsible for about one in four deaths worldwide. **Pneumonia**, where the alveoli become filled with fluid, is responsible for the majority (43%) of all deaths from respiratory diseases – it is a frequent cause of death in the elderly. This is followed by **respiratory cancers** (23%) and a set of diseases: asthma, chronic bronchitis and emphysema (together responsible for 20%). The latter two conditions may lead to similar irreversible symptoms which are referred to as chronic obstructive pulmonary disease.

3.3.1 Asthma

Asthma is one of the fastest-growing health problems throughout the world. Since 1980, the number of people affected worldwide has more than doubled. Currently, asthma affects about 3.5 million people in Britain, 1.4 million of whom are children (3–15 years). The onset of asthma normally occurs in early childhood and is responsible for nearly one-third of all long-term illnesses in children.

Asthma is associated with a narrowing or constriction of the airways, swelling of the airway linings, an over-production of mucus and difficulties in expiration. There are over 200 known factors that trigger asthma. These include pollen, house dust mite, animal fur contaminated with urine, pollutants, dairy produce, chest infections, exercise, emotional stress and other factors, any of which may trigger an *allergic reaction*. When acute asthma attacks occur, the first line of drugs are **bronchodilators**, delivered as an aerosol by inhalation from small pressurized canisters. The commonly prescribed drugs, *salbutamol* or *salmeterol*, stimulate sympathetic β2-adrenoceptors in bronchial smooth muscle to cause bronchial dilation and a decrease in airway resistance (these drugs are called β2-adrenoceptor *agonists*). In severe acute asthma attacks, additional drugs such as *iprotropium bromide*, that block the bronchoconstrictive action of parasympathetic acetylcholine are also administered. The effect of these bronchodilator drugs is to increase air flow through the bronchioles, which greatly eases the effort of expiration and relieves the symptoms of asthma.

Second-line corticosteroid drugs, such as inhaled *beclomethasone*, are used as regular long-term preventative medication – called **prophylactic agents**. These drugs act by reducing bronchial mucosal inflammatory reactions (e.g. oedema production and excess mucus secretion) and by modifying cellular reactions to potential allergens. Sudden acute asthma attacks – particularly those in the young – that are not controlled by a patient's prescribed drugs should be treated as a medical emergency requiring hospitalization; if left unchecked, these attacks can worsen swiftly and become fatal due to severe respiratory problems (see Case Report 3.1).

Case Report 3.1   A life with asthma

Caroline is now 19 years old and has had asthma and eczema since she was a young child. Between the ages of 3 and 5 years, she had three acute admissions to hospital after becoming extremely asthmatic. Her mother was so worried by these events that she joined the National Asthma Campaign and set up a local group to provide support for other families with asthmatic children.

During Caroline's childhood, her condition was well controlled with inhalers. Like many children, she had a corticosteroid *(beclomethasone)* inhaler for prevention and a bronchodilator inhaler (*salbutamol*) which she used if necessary. Her mother monitored her condition using a peak flow meter. Between the ages of 5 and 14 years, there was only one occasion when her asthma was so serious that she needed to be taken to the Accident and Emergency (A&E) Unit of her local hospital. Whilst staying with her cousins, she had become very wheezy, despite using her salbutamol inhaler. At hospital, she responded quickly to bronchodilators (*salbutamol* and *iprotropium bromide*) administered via inhalation and was soon discharged.

However, her mother remembers an occasion when Caroline (then 15) had became so wheezy that an emergency appointment was made with her GP, who immediately called an ambulance. In hospital, her condition was so serious that even nebulized bronchodilators and intravenous corticosteroids (to reduce inflammation in the lungs) had no effect. Only an intravenous infusion of *aminophylline* (a potent bronchodilator) was sufficient to manage her acute condition. Nurses monitored Caroline closely during the infusion to ensure her heart rate and blood pressure did not increase dramatically – a possible side-effect of bronchodilators. Caroline recovered fully and left hospital. As part of managing her asthma, she was told to check her peak flow rate each day and use her corticosteroid inhaler on a daily basis. By this time, Caroline was found to be highly allergic to house mites and cats and was advised about the effect of these allergens on her asthma.

As Caroline's social life developed, she paid less attention to the daily management of her asthma. Her mother became very concerned because some of Caroline's friends smoked, possibly making Caroline's asthma worse. Several of Caroline's friends had furry cats. Since Caroline enjoyed her social life, she should have coped with the reality of her asthma by making sure she had her salbutamol inhaler with her at all times and using it at an early stage. She should have been taking her corticosteroid inhaler daily and monitoring her peak flow regularly.

In her first term at university, Caroline was with friends in a bar when her breathing suddenly became very difficult. Unfortunately, her salbutamol inhaler was empty and she didn't have her spare one. Her friends noticed how laboured her breathing was and immediately rushed her to the nearest A&E. Her pulse rate was found to be significantly raised, at 104 beats per minute, her peak flow rate was 280 litres per minute (about half her normal value) and her blood-oxygen saturation level was down to 90% (normal is about 99%). Caroline was given oxygen to increase her oxygen saturation and a nebulized dose of *salbutamol* and *iprotropium bromide* every 20 minutes to increase bronchodilation and peak flow, thus alleviating the symptoms. Caroline was reluctant to take the prescribed oral corticosteroids – she was worried they would make her fat. Thankfully, Caroline's condition improved and she was discharged from hospital.

Caroline's story is not unique – it is likely that she, like many others who have had asthma since childhood, will have to cope with asthma for the rest of her life.

● In Case Report 3.1, why did the nurses closely monitor Caroline's heart rate and blood pressure during the intravenous infusion of the potent bronchodilator aminophylline (a β2-adrenoceptor agonist)?

● Aminophylline stimulates not only β2-adrenoceptors in the lungs (producing bronchodilation and alleviating the symptoms of asthma) but also stimulates the β2-adrenoceptors in the heart. In individuals who are particularly sensitive to aminophylline, this could produce an increase in heart rate and force of cardiac muscle contraction, leading to subsequent hypertensive effects. These potential side-effects need to be carefully monitored. (However, at the clinical doses that cause bronchodilation, possible side-effects of heart stimulation do not usually occur.)

3.3.2 Bronchitis

On average, we breathe about 20 000 times a day, taking approximately 16 kg or 10 000 litres of air into our lungs. Any pollutants or small airborne particles present in the air (such as nitrogen oxides, carbon monoxide, sulfur oxides, dust or cigarette smoke) will also be inhaled. As mentioned earlier, the respiratory system works hard to remove any noxious particles or allergens by means of the cilia and mucus produced by the cells lining the large airways. Consequently, the bronchioles may become sore and inflamed, leading to bronchitis. With prolonged exposure to pollutants, the system responds by narrowing (constricting) the airways, further reducing the volume of contaminated air entering the lungs and passing through to the alveoli. The airways also respond by producing excessive mucus. The condition can progress so that the airways become severely inflamed, leaving them damaged and highly infected – a condition known as **chronic bronchitis**.

● Why do you think that people with chronic bronchitis find it difficult to perform physical exercise?

● Due to bronchiolar constriction and mucus production, the gas exchange capacity of the lung will be less efficient. As a result, the concentration of oxygen entering the blood will be greatly reduced to an extent that may be insufficient to support the metabolic demands of muscular exercise. Breathing rate would need to be increased to a point that may be impossible to sustain.

3.3.3 Emphysema

Emphysema (Greek: *emphusema* means 'to inflate') is a lung disease in which the elastic supporting structure between alveoli has been irreversibly damaged. This leads to the breakdown and collapse of numerous air sacs and larger airways, with the result that air becomes trapped and the lungs 'over-inflated'. Emphysema generally occurs as a result of smoking and leads to severe breathlessness and wheezing. Here, the respiratory system has to work extra hard to obtain sufficient O_2 for the body's metabolic demands, because of the greatly reduced gas exchange potential of the lungs. To compensate, cardiac output increases, so increasing the volume of blood pumped to the lungs, and the right ventricle may become abnormally enlarged (hypertrophy). The end result may be attacks of myocardial ischaemia, ventricular failure, and ultimately, heart attack.

3.3.4 Chronic obstructive pulmonary disease

Chronic obstructive pulmonary disease (COPD) encompasses a number of overlapping conditions – principally chronic bronchitis and emphysema, but may also involve asthma. The common causes of COPD are smoking, industrial pollutants (e.g. dust particles, smoke) and possibly genetic factors. In severe cases, lung function is so profoundly compromised that patients are virtually sedentary and are given **long-term oxygen therapy**, using a face-mask or nasal cannula (tube inserted into the nose) to deliver O_2 from a gas cylinder, to assist with the reduced respiratory capacity of their diseased lungs. Unfortunately, there are currently no cures for COPD and once started it gets worse, drastically affecting the lifestyle of the sufferer (see Case Report 3.2).

Case Report 3.2   Living with chronic obstructive pulmonary disease

Michael is 70 years old and lives with his wife in a three-bedroomed terraced house. He has grown-up children, who live locally, and he and his wife greatly enjoy contact with their grandchildren. Michael was diagnosed with emphysema more than 20 years ago. Although his life has become increasingly restricted, he has maintained a reasonable level of activity for much of the time. He now has continuous oxygen therapy and can only leave his home for short periods. What follows is his own account of how his illness developed and how it has affected him and his family.

It was about 1980, I first noticed that if I ran upstairs, by the time I got to the top I was really breathless. I used to smoke about 50 cigarettes a day and I was always in the pub or a club in London or somewhere. Then I noticed that I'd be in the pub just talking away and I'd just lose my breath for no reason. I saw my GP and eventually in about 1984, I was told that I had emphysema – I gave up cigarettes straight away. But about six months later I ended up in hospital because I could not breathe. I was on a ventilator for about 10 days. When I was about to be discharged, I went to the bathroom and had a wash – but this made me totally breathless. I said to the consultant, "Actually doctor, I don't think I'm well enough to go home because I still can't breathe, I can't do anything." He said to me, "Mr Read, you've got emphysema. I'm afraid your condition isn't going to get better – it's likely to gradually get worse, and you could get problems with your heart too." So I thought, good grief, it's really serious this – far more serious than I thought.

So I came home, and it was quite a shock to the family that my condition wasn't going to get better, but it wasn't affecting me too much. I could do things very slowly. I didn't need oxygen but I needed an inhaler and over the years I gradually got worse. My two boys wanted apprenticeships as carpenters and joiners, which I served my apprenticeship in, so I said, "As long as you can do the physical stuff, I'll teach you. We'll take on a few jobs, and I'll teach you what to do." So over the next five years, that's what we did. I did all the designing and they did the physical stuff. Every six months, I used to go and see the consultant, and about five years ago he said that I needed to be on continuous oxygen, as my oxygen levels were so low. He said it should be for a minimum of 16 hours a day, and that it would prolong my life. So I'm all set up with oxygen equipment at home and now I can't be without it. If I haven't got my oxygen with me and I get breathless, I panic.

About three years ago, one night I just couldn't get my breath. I asked my wife to ring for an ambulance and it turned out I had pneumonia. I was in hospital for about ten days. I always cough up sputum and the nurses explained that I should check the colour of it as if it changed colour – green in particular this was a sign of infection and I should see my GP. Eventually I did get home and I got a wheelchair and a stair lift to help me get upstairs. Since then, I've recovered and I'm fairly normal except that I have to have oxygen 24 hours a day and I can only walk around the house. I have [salbutamol and corticosteroid] inhalers that I take four times a day. If I go out, it has to be in a wheelchair and I have to take my portable oxygen cylinder with me.

This is graphic account of how emphysema can affect the quality of a person's life and the lives of those around. Although the disease is irreversible, Michael is philosophical about the future, 'I'm fairly normal except…

Having read about Michael's medical history in Case Report 3.2, take a few moments to imagine yourself in his place.

Every breath you take is a painful laboured effort, the slightest exercise an aerobic challenge. Your whole life revolves around what you used to take for granted – trouble-free inspiration and expiration. Walking around the house is now an exhausting marathon. You have several inhalers to take throughout the day. Furthermore, an oxygen mask must be worn 24 hours a day, 365 days a year for the rest of your life. If you go out, it has to be in a wheelchair equipped with a portable oxygen cylinder. You will always need someone to push your wheelchair. Night-times are difficult. Relationships with family, friends and neighbours have entered new levels of physical and emotional involvement. This is your reality – a reality that shapes every moment of your life experience.

The reason for asking you to think about this scenario, and to empathize with Michael, is to illustrate how respiratory disease can affect the whole person – as well as their family and friends. Moreover, it isn't a static occurrence, it is a developmental phenomenon that ranges widely from the molecular pharmacology of lung function right through to the psychology of patient–carer interaction and beyond (see Book 1, Sections 1.3 and 1.7).

The British Thoracic Society website contains useful additional information about lung diseases including COPD (see Further Reading).

3.3.5 Care for a cigarette?

The days of the large advertising posters showing the cigarette touting Marlboro cowboys and enticingly cool mountain streams are long gone. Gone too are the long relaxing after-dinner cigarettes and smoking in confined communal areas. Did you know that by inhaling the smoke from a lit cigarette, over 4000 different chemicals (including ammonia, phenols, naphthalene, DDT insecticide, cadmium, carbon monoxide (CO), arsenic and tiny droplets of tar and nicotine) could directly enter your lungs?

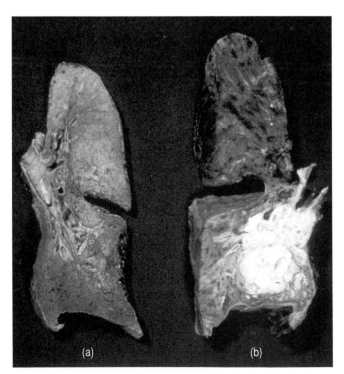

Figure 3.9 Lung specimens from (a) a non-smoker and (b) a smoker. Note that the smoker's lung is nearly black with tar deposits, misshapen and the surface is very rough in texture and riddled with cancer (white areas).

Conclusive medical and epidemiological evidence now indicates that smoking extensively damages both lung structure and function, and seriously compromises general health (Figure 3.9). Heavy smokers have abnormally raised levels of CO and carboxyhaemoglobin in their blood, leading to a greatly reduced capacity to transport O_2. In addition, constant irritation by the tar in smoke causes chronic bronchitis, where the air passages in the lungs become intensely and repeatedly inflamed; this often leads to emphysema. The walls of these airways thicken, leading to the build-up of fibrous scar tissue, which greatly reduces the elasticity and efficiency of lung function. Lung cancer, which is caused by both nicotine and tar, leads to uncontrollable tumours that frequently destroy lung tissue and can severely compromise health and life expectancy.

Lung cancer is now the most common form of cancer diagnosed in the USA and UK and is a major cause of death. Lung cancer accounts for 14% of all deaths and 28% of all cancer deaths. Cigarette smoking is responsible for an estimated 87% of all lung cancer deaths. Put another way, seven out of eight lung cancer deaths are related to smoking. Care for a cigarette?

It might surprise you to know that epidemiological evidence was being used over 500 years ago. The German philosopher and medical reformer Paracelsus (1493–1541) was one of the first 'scientific physicians' to make a detailed catalogue of lung disease. In his monograph entitled *On the Miner's Sickness*, he identified lung-related ailments, especially *silicosis*, as a direct result of working in the mines and the subsequent effects on other physiological systems in the body, including heart-related problems. Such research greatly emphasized the central importance of lung function in the overall health of the body.

Summary of Section 3.3

1   Respiratory diseases account for 24% of all deaths in the UK. Taken together, asthma, chronic bronchitis and emphysema account for one-fifth of these deaths.

2   Asthma is a very common respiratory condition caused by many factors ranging from allergens to emotional stress. It is generally associated with a narrowing of the airways, swelling of the airway linings, over-production of mucus and difficulties in expiration. These symptoms are largely prevented by the prophylactic use of corticosteroids and quickly relieved by the direct inhalation of bronchodilators.

3   Pollutants and particles in inspired air can cause inflammation of the bronchioles – a disease called bronchitis. Prolonged exposure to these agents leads to chronic bronchitis. This results in excessive bronchiolar constriction and mucus production, which significantly reduces the gas exchange capacity of the lungs.

4   Emphysema is a progressive irreversible disease resulting in the widespread collapse of the alveolar walls in the lungs, with the result that air becomes trapped and the lungs 'over-inflated'. A major cause of emphysema is smoking.

5   Chronic obstructive pulmonary disease (COPD) is also a progressive, irreversible lung disease with causes and symptoms similar to chronic bronchitis, emphysema and asthma. Breathing becomes so severely compromised that patients may need supplementary oxygen.

6   Smoking can seriously damage health. The great majority of deaths from lung cancer relate to cigarette smoking.

3.4 Cardiorespiratory integration – heart and lungs act in concert

Although we have mentioned some of the control mechanisms of both the circulatory and respiratory systems individually, it should be apparent that the workings of each are closely interlinked. An increase in the O_2 requirements of the tissues, e.g. during exercise, will lead to an increase in heart rate and a simultaneous increase in ventilation rate. Many of the sensors that monitor these changes, such as the baroreceptors and chemoreceptors, lead to an adjustment of the activity of both systems at the same time. In this section, we will look in greater detail at the control of both the respiratory and cardiovascular systems and consider them as one integrated system. Figure 3.10 (overleaf) shows some of the interactions between the two systems.

Figure 3.10 An illustration of the interactions between the cardiovascular and respiratory systems leading to alterations in cardiac output.

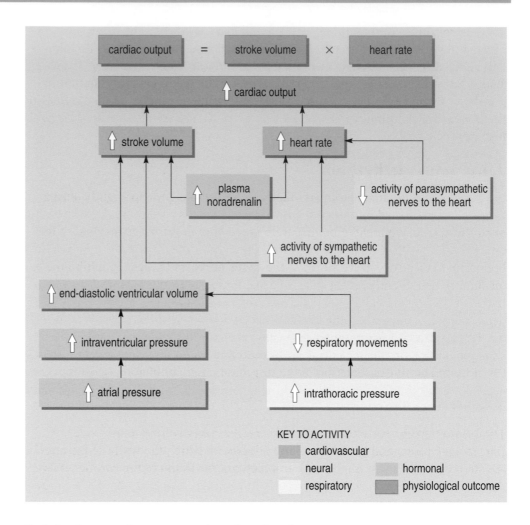

3.4.1 Control centres in the brain

Although the heart rate and rate of breathing can be quickly modified by changes in the internal or external environment, the overall control of both systems is driven by neural activity from the brain. This section is a brief review of the neural centres in the brain controlling the cardiovasular and respiratory systems. The control centres are primarily located in the medulla and form part of the brainstem. If there was no centre controlling the heart rate, the heart would contract at the rate of activity of the SAN, i.e. at about 80–90 beats per minute. The SAN receives parasympathetic inputs from the vagus nerve. Activity in the vagus nerve input to the SAN reduces the heart rate, an effect called 'vagal restraint' (Section 2.4.6). This tonic inhibition of the SAN from the parasympathetic vagus nerve reduces the intrinsic heart rate to, on average, 70 beats per minute. The tonic inhibition is initiated in the cardio-inhibitory centre in the medulla.

- Why is it necessary to have tonic inhibition to reduce the heart rate below that of the SAN?

- If the heart was beating continually at the full potential of the SAN, it would not be possible to increase the heart rate when required. A reduction in tonic inhibition from the cardioinhibitory centre in the brain will increase heart rate.

The control centre for breathing is also found in the medulla. The inspiratory centre drives the inspiratory nerves, stimulating the activity of the inspiratory muscles and at the same time inhibiting the expiratory centre. The respiratory centre responds to the activity of stretch receptors in the lungs by (a) inhibiting activity of the inspiratory nerves and thus terminating inspiration, and (b) releasing the block on the expiratory nerves. Activation of the latter causes relaxation of the inspiratory muscles.

3.4.2 Sinus arrhythmia

In a healthy individual breathing normally, the heart rate goes up slightly during inspiration and goes down slightly with expiration. This variation is known as **sinus arrhythmia**. As we breathe in, the stretch receptors in the lungs respond to the expansion of the lungs. This information is relayed to the medullary region of the brainstem via sensory nerve fibres in the vagus nerve and serves to inhibit the inspiratory nerves and prevent excessive inspiration. This inhibition is also relayed to the cardiac centre and the tonic or continuous vagal activity (which slows the heart rate under normal conditions) is reduced, leading to an increase in activity of the SAN and therefore an increase in heart rate. During expiration, the reverse situation occurs, producing a decrease in heart rate. Sinus arrhythmia demonstrates the close integrative association of the respiratory and circulatory systems

3.4.3 The baroreceptors and chemoreceptors

The location of the baroreceptors and chemoreceptors is shown in Figure 3.11. Baroreceptors respond to changes in the degree to which the walls of the arteries are stretched. The most important baroreceptors are found in the carotid sinus (Section 2.4.6) of each of the internal carotid arteries which supply the brain. A second set of baroreceptors are found in the walls of the aorta at the aortic arch. Also in this area are a number of stretch receptors which respond to the distension of the aorta that results from an increase in blood pressure. The chemoreceptors are found in small extensions of the aorta and in the external carotid arteries and are known as aortic and carotid bodies respectively. Information from all these receptors is delivered to the medulla of the brain by specific sensory fibres in the vagus and glossopharyngeal nerves (Figure 3.11). The activity in these nerve fibres is tonic, i.e. they fire at a slow rate under 'normal' conditions, but if the conditions change they can increase or decrease their rate of firing accordingly.

- The vagus nerve (10th cranial nerve) is involved in many aspects of cardiovascular and pulmonary function. But do you know what other structures it innervates?

- The vagus nerve innervates not only the heart and lungs but also the kidney. It also innervates the gall bladder and the gastrointestinal tract – a 'wandering' vagrant nerve indeed!

Thus, there are 'loops' of neural activity between the baroreceptors and chemoreceptors, the medulla of the brain (the area where the respiratory and cardiac centres are found), and the effectors (the sinoatrial node of the heart and the inspiratory muscles). These are examples of reflex 'homeostatic feedback mechanisms' (see Book 1, Section 2.4).

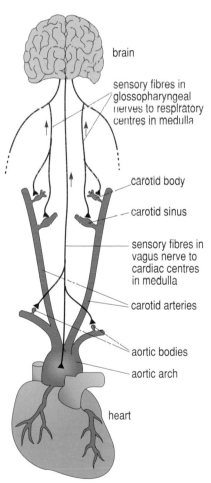

brain

sensory fibres in glossopharyngeal nerves to respiratory centres in medulla

carotid body

carotid sinus

sensory fibres in vagus nerve to cardiac centres in medulla

carotid arteries

aortic bodies

aortic arch

heart

Figure 3.11 The location of the aortic arch, aortic bodies, carotid sinuses, carotid bodies and the associated afferent (sensory) nerves.

What happens if the blood pressure changes? If the arterial blood pressure increases, the reflex feedback will bring into effect changes that reduce it to normal values. The increase in blood pressure causes the walls of the arteries to be stretched and the change is sensed by the baroreceptors and stretch receptors. This increases the rate of firing of the nerve fibres from the carotid sinus and aortic arch. The information is relayed to the cardiac centre, which increases the vagal (parasympathetic) activity to the heart, so reducing the heart rate. There is also a reduction in the activity of the sympathetic nerves. This reduces the stroke volume and causes the smooth muscle in the arterioles to relax, leading to vasodilation. Overall, the reduction in heart rate and stroke volume reduces cardiac output, and vasodilation reduces the peripheral resistance. All these changes together reduce the arterial blood pressure. An increase in the activity of the baroreceptor nerves leads to a small reduction in inspiration, although this has very little physiological benefit. Figure 3.12 shows schematically the reflex feedback in response to an increase in arterial blood pressure.

● From Figure 3.12, describe what will happen if the baroreceptors detect a decrease in blood pressure.

● A reflex increase in blood pressure would occur. This would result from a reduction in the firing rate of the nerves from the baroreceptors to the cardiac and respiratory centres, causing a decrease in parasympathetic activity of the vagus nerve to the heart, and hence an increase in heart rate. Stroke volume would be increased and vasoconstriction would occur due to an increase in the firing rate of sympathetic nerves. Cardiac output and total peripheral resistance would increase, thus increasing arterial blood pressure.

What happens if there is a change in the partial pressure of O_2 in the arterial blood? The chemoreceptors in the carotid and aortic bodies are the only chemoreceptors that respond to changes in P_{O_2}. Other chemoreceptors, like those in the respiratory centre of the brain (see Section 3.2.6), respond to changes in H^+ ion concentration as a result of changes in P_{CO_2}. If the P_{O_2} is reduced due to an increase in the rate of utilization of O_2 by the tissues or due to a decrease in the blood flow through the carotid and aortic bodies, a reflex feedback will be initiated to return the P_{O_2} value to normal. The activity in the nerve fibres from the chemoreceptors in the carotid and aortic bodies increases in response to a fall in the arterial P_{O_2}. This information is relayed to the respiratory and cardiac centres in the medulla and a reflex increase in the rate and depth of breathing is initiated by stimulation of the inspiratory neurons. This serves to increase the P_{O_2} in the alveoli. At the same time, the heart rate, stroke volume and total peripheral resistance are increased, which in turn raises the arterial blood pressure and increases the flow of blood through the carotid and aortic bodies. These chemoreceptors respond in a similar manner to an increase in arterial P_{CO_2} or H^+ ion concentration. A point to note is that the respiratory system is more sensitive (i.e. responds more rapidly) to a change in H^+ concentration than to a change in O_2 concentration.

3.4.4 When things go wrong...

All the changes that have been described above are reflex changes which are initiated subconsciously. In the next section, we will discuss a few situations where the cardiovascular and respiratory systems respond to changes in the environment,

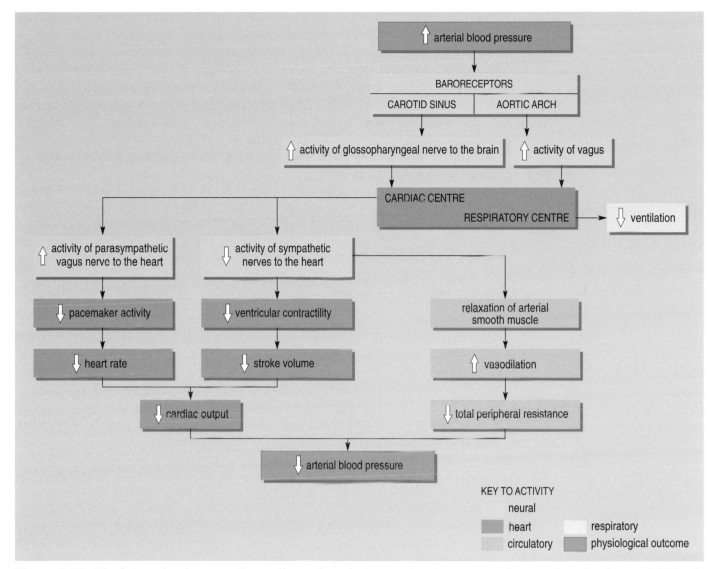

Figure 3.12 The interaction between the cardiac and respiratory centres to accommodate an alteration in arterial blood pressure.

and the responses in some cases are voluntary. The maintenance of a fully functional, integrated cardiovascular and respiratory system is dependent on the continual accurate monitoring of the circulating blood by chemoreceptors and baroreceptors. If either of these sensor systems fails, the reflex feedback mechanisms will not maintain the homeostatic control of heart and lung function.

When we are asleep, the control of respiration is not as rigorous as when we are awake. Consequently there are times during sleep when we stop breathing briefly. These periods are known as **apnoea** and are due to a decrease in the sensitivity of the chemoreceptors to alterations in P_{CO_2}. For some individuals, these bouts of apnoea become a serious problem; they wake up with headaches, fatigued, and in the most serious cases show all the clinical symptoms of respiratory failure although they have perfectly healthy lungs. This is known as 'sleep apnoea syndrome' and one of the causes is believed to be a failure, during sleep, of contraction of the muscles that normally hold the tongue in place. The tongue drops back, blocking the airway and preventing air entering the lungs (see Figure 3.1). There is a consequent rise in blood P_{CO_2} but this may not be detected if there is a decreased sensitivity of the chemoreceptors. As a result there may be a failure to instigate the reflexive mechanisms responsible for an increase in ventilation – the consequences may be disastrous. Some instances of **sudden infant death syndrome** (**SIDS**), more commonly known as cot death, may be caused by a form of sleep apnoea, due to a similar malfunctioning of the chemoreceptors. There is still much debate about the cause of this strange and tragic phenomenon.

3.4.5 Aspects of emotional and voluntary influences on breathing

Involuntary modifications of cardiorespiratory function occur during the expression of various emotional states such as laughter, crying, sighing and groaning. For example, natural laughter and the associated physiological and psychological activities are now known to lower blood pressure, to stimulate the immune system, favourably alter mood and reduce stress – indicating that humour can be good for your health (Hassed, 2001).

Humans also have considerable voluntary control over breathing. This is accomplished by activity in parts of the cerebral cortex, such as the frontal lobes, involved in conscious, deliberate action. These brain regions are known to strongly influence the cardiorespiratory centres in the medulla and therefore have the ability to alter – sometimes dramatically – the pattern of breathing. For example, we can voluntarily hyperventilate ('over-breathe') or hold our breath.

How long can you hold your breath? …30, 45 seconds or a minute? Amazingly, the world record breath hold is over eight minutes (Phillips, 2001).

- In what other daily activities do we consciously influence breathing?

- We deliberately alter the pattern of ventilation during such everyday activities as speaking, singing, whistling or playing a woodwind or brass musical instrument. Some meditative practices require the pattern of breathing to be consciously altered (e.g. pranayama yoga). Moreover, breathing can also be used to control emotions – for example, taking regular, deep breaths when nervous or angry.

Summary of Section 3.4

1 The circulatory and respiratory systems work in unison to maintain blood flow and respiration.

2 The close interaction between the cardiovascular and respiratory systems is shown by the phenomenon of sinus arrhythmia, where the heart rate increases as we breathe in and decreases as we breathe out.

3 The baroreceptors and chemoreceptors are central to the integrated control of both systems and initiate reflex, subconscious changes in blood pressure and respiration rate.

4 If the baroreceptors or the chemoreceptors fail to initiate these reflex changes, then problems of respiration and blood flow result.

5 Emotions and conscious thought can greatly influence cardiorespiratory function.

3.5 Cardiorespiratory adaptation – meeting different demands

In Section 3.4, the neural control of the cardiovascular and respiratory systems was described. Here, we will look at a number of conditions requiring rather more drastic alterations in cardiovascular and respiratory physiology. It should be noted that although these modifications may appear to be more of a conscious alteration in the functioning of the two systems, they are still stimulated, in the main, by the involuntary reflexes described above.

3.5.1 Adjustments during exercise

The cardiovascular and respiratory adjustments that take place to sustain us through exercise actually begin before exercise commences, i.e. there are anticipatory changes preparing the body for the increased demands to be made on its resources. The first anticipatory event is activation of the sympathetic nervous system and the release of adrenalin from the adrenal glands into the bloodstream. This leads to an increase in the strength of contraction of the ventricles. Simultaneously, vasoconstriction of the arterioles increases the arterial blood pressure and vasoconstriction of the venous system increases venous return, which increases stroke volume and cardiac output. The onset of exercise is accompanied by vasodilation of the muscle blood vessels, which increases blood flow through the muscles. As the muscles continue to exercise, the metabolic demands increase, i.e. the cells of the active muscles need an increased supply of glucose and O_2 for respiration. This increase in metabolic rate is the most significant factor in the local control of blood flow through the exercising tissues. Carbon dioxide is produced in greater quantities than when at rest and acts as a local vasodilator, increasing the blood flow though the tissues. The increase in temperature and the fall in pH (due to the increase in CO_2) causes the Bohr shift in the oxygen–haemoglobin dissociation curve, and so O_2 is released from oxyhaemoglobin at higher partial pressures of the gas, i.e. O_2 is transferred to the tissue cells more readily.

In anticipation of exercise, the ventilation rate is also increased in response to raised levels of adrenalin and noradrenalin activating the inspiratory muscles. The increased rate of ventilation is maintained during exercise due to the high demands for O_2.

● By what mechanism is the increased demand for O_2 in the body detected?

● The decrease in P_{O_2} and the increase in P_{CO_2} in the blood, following the increased use of O_2 by exercising muscles, is detected by both central and peripheral chemoreceptors.

● What effect does this have on the rate of ventilation?

● The rate of ventilation is increased. The chemoreceptors effect an increase in the efferent (motor) activity of the inspiratory nerves, which stimulates inspiration.

Stretch receptors in the muscles are also activated during exercise and they inform the respiratory centres of the need for increased respiration. As exercise continues, the increased demands for O_2 and glucose cannot always be met by increased ventilation and increased blood flow through the tissues. The muscle cells begin to respire *anaerobically* (i.e. without consuming O_2), and mobilize energy sources such as glycogen, a storage carbohydrate. Although anaerobic respiration can provide energy for the muscles, one of the by-products of respiration in the absence of O_2 is lactic acid or lactate (see Book 2, Section 4.8.4). The excess lactate builds up in muscle and can only be removed when sufficient O_2 is again delivered to the muscles.

● How does a build-up of lactate lead to an increase in O_2 delivery?

● An increase in acid concentrations in the muscles will shift the oxygen–haemoglobin dissociation curve to the right and this will increase the release of O_2 from oxyhaemoglobin.

As you learned in Book 2, Section 4.8.4, most of the lactate produced during exercise is transported in the bloodstream to the liver where it is metabolized to replenish glycogen stores. Some lactate is broken down in the muscle tissue to yield energy. Both of these processes, however, require oxygen and, although the muscles are no longer exercising, the rate of O_2 supply must remain elevated until all the lactate has been removed or broken down. The O_2 needed after exercise to metabolize lactate is known as the excess post-exercise oxygen consumption (EPOC). The build-up of lactate in the muscles is often experienced as a type of cramp. This can be painful and is only relieved when the lactate levels have fallen. When exercise is finished, a continued increase in ventilation provides the necessary O_2. The heart rate remains elevated as long as the ventilation rate is increased and this maintains the flow of blood through the muscles, which removes the lactate and delivers the necessary oxygen. During sustained exercise, the initially high O_2 requirement can be at least partially met while the muscles are still working. An equilibrium may be established between the supply and demand for O_2 once the exercise rate has reached a so-called *steady state*. The period following this establishment of equilibrium is known as the 'second wind'. However, its onset depends on the type and level of exercise and the fitness of the individual; even world-class endurance athletes cannot perform beyond their maximum lactate tolerance.

The increases in ventilation and cardiac output which allow the muscles to exercise are brought about by a combination of neural activity and responses to local changes in the chemical composition of the blood. The anticipatory changes in cardiac output and the rate of ventilation set the background that permits sustained exercise.

How much O_2 do you consume doing physical work?

The best indicator of a person's ability to undertake aerobic work is their maximal O_2 consumption (or max. V_{O_2}). This is the maximum volume of O_2 an individual is capable of using per minute to oxidize nutrient molecules for energy production. Max. V_{O_2} is measured by getting a person to exercise on a treadmill or a bicycle ergometer. The exercise workload is gradually increased until the person approaches exhaustion. During the latter stages of exercise, when O_2 consumption is very near maximal, the percentage of O_2 and CO_2 in expired air samples is determined. The volume of the expired air is also measured. Physiological equations (which take into account the percentages of O_2 and CO_2 in the inspired air, the total volume of air expired, and the percentages of O_2 and CO_2 in the expired air) are then used to calculate the amount of O_2 consumed at near-maximal exercise – this is finally expressed as litres of O_2 per minute. So that individuals with different body weights can be compared, this value is expressed as millilitres of O_2 consumed per kilogram of body weight per minute ($ml\ O_2\ kg^{-1}\ min^{-1}$).

● What are the major physiological variables that affect maximal O_2 consumption (or max. V_{O_2})?

● The five main physiological variables that influence an individuals max. V_{O_2} are:

- cardiac output (= heart rate × stroke volume);
- the functional capacity of the lungs;
- the efficient exchange of O_2 and CO_2 between the air and the blood in the lungs (mediated by the respiratory system);
- the efficient delivery of oxygenated blood to working muscles (the performance of the cardiovascular system);
- the efficient operation of the biochemical pathways underlying aerobic activity in muscle cells.

Regular aerobic exercise can improve the performance of all these variables. As expected, Tour de France cyclists have a very high max. V_{O_2} of $85–100\ ml\ O_2\ kg^{-1}\ min^{-1}$, Olympic athletes have values of $60–85\ ml\ O_2\ kg^{-1}\ min^{-1}$, whereas sedentary people consume about $25–40\ ml\ O_2\ kg^{-1}\ min^{-1}$ (while at rest).

Exercise physiologists have calculated the average aerobic capacity for different age, gender and weight groups, so that exercise training regimes can be designed for people to achieve their optimal level of aerobic fitness.

Do you do any regular aerobic exercise? A wealth of evidence from a variety of medical, psychological and sociological disciplines indicates that regular aerobic exercise is good for you. The benefits range from improved cardiovascular and respiratory fitness, increased joint mobility and better physical health, to weight and diet control, improved sleep patterns and also psychological benefits; there are two that are particularly important:

- Exercise can greatly alleviate the effects of depression and stress (Chapter 3 in Book 4 discusses stress in detail).

- Regular aerobic exercise may be beneficial in delaying the effects of neurodegenerative diseases; exercise enhances the performance of the frontal areas of the cerebral cortex in the brain that are known to be selectively affected during the progression of dementia, schizophrenia and Parkinson's disease (Sutoo and Akiyama, 2003).

3.5.2 Adjustment to high altitude

The partial pressure of O_2 in the atmosphere decreases with increasing altitude. Indeed people commonly talk about the 'atmosphere thinning' when ascending a mountain. The ability to adjust to the decrease in P_{O_2} depends on the altitude reached and the speed of ascent. For example, if a pilot flies an aircraft directly up to 4000 m without supplementary oxygen, symptoms of hypoxia will rapidly develop, including headache, drowsiness, impaired judgement, loss of pain sensations, excitement, disorientation and loss of sense of time. If the pilot carries on ascending to 8000 m, unconsciousness will eventually occur. If, however, the pilot makes the full ascent to 8000 m slowly, and is provided with supplementary O_2 (delivered via an oxygen mask), these symptoms do not occur.

Similarly, if a mountaineer gradually ascends a mountain over a period of weeks, they will eventually become adapted to the low levels of O_2 in the air. However, at around 4000 m some mountaineers experience *mountain sickness*, a form of hypoxia, where they suffer headaches, feel nauseous and can become easily confused during tasks that require mental concentration and thought. As they remain at that level for a few weeks, the symptoms wear off and they become physiologically adapted. So what cardiopulmonary adjustments occur to allow this adaptation to occur? There are five main compensatory events which permit adaptation to high altitude:

(i) increased pulmonary ventilation in response to low P_{O_2} and high P_{CO_2} detected by the peripheral and central chemoreceptors;

(ii) an increase in the carrying capacity of the blood for O_2, due to an increase in the number of erythrocytes and an increase in haemoglobin production (i.e. both more erythrocytes and more haemoglobin per cell);

(iii) an increase in the blood volume, which increases cardiac output and the amount of blood flowing in the capillaries of the lungs, thus allowing more O_2 to diffuse from the alveoli to the blood;

(iv) an increase in blood supply to the tissues because there is an increase in the number of blood vessels in the tissues, i.e. the vascularization of the tissues is increased;

(v) an adaptive increase in the ability of the tissues to utilize O_2 at higher P_{CO_2}.

● Can you suggest why athletes choose to train at high altitude to improve their cardiovascular fitness and their pulmonary ventilation?

● After a period of time spent training at high altitude, the body adjusts and can function normally at low P_{O_2}. Oxygen-carrying capacity is increased and the tissues utilize O_2 more efficiently. Cardiac output is increased, ventilation is maximized, and there is an increase in the number of erythrocytes. On return to lower altitude, these compensatory adjustments are retained for a short period, allowing athletes to perform at a higher intensity than they did previously at low altitude.

3.5.3 Adjustments during diving

Deep-sea diving poses a further problem for the respiratory and cardiovascular systems. As well as the requirement for increases in ventilation and cardiac output sufficient to sustain exercising muscles, the body has to cope with the problem posed by increased water pressure in the diving environment. As the body goes deeper under water, the pressure of water exerted on the body increases. At some critical point, this extreme external pressure will cause the lungs to collapse. Deep-sea divers therefore have to breathe in air that is under higher pressure than the water, i.e. compressed air, to prevent the lungs collapsing as they dive. There is, however, a problem when breathing compressed air. As well as oxygen, the nitrogen in compressed air is forced through the alveoli, into the blood and into the tissues, particularly lipid-rich ones. This is not a problem under water, because there is a sufficient supply of O_2 to the tissues. However, when the diver comes back up to the surface, the water pressure is reduced, but the pressure of the gases inside the body are still equal to those of the compressed gas, i.e. much higher than the surrounding pressure. The gases begin to come out of solution and can form gas bubbles in the blood and tissues: since nitrogen makes up nearly 80% of the inspired gases, it is the major cause of problems. Bubbles of nitrogen formed in the body cause *decompression sickness*, commonly known as *the bends*. In the most serious cases of the bends, bubbles of gas coming out of solution can tear through tissues or block blood vessels, and damage nerve pathways in the brain and spinal cord; they can cause severe pain and may lead to permanent paralysis, permanent abnormal mental function or, in severe cases, death.

The bends can be avoided if the diver is brought to the surface very slowly so that the gases do not come out of solution while still in the tissues and are eliminated from the body through the lungs. For example, if a diver descends to a depth of 65 m below the surface and remains there for two hours, then the time needed to bring the diver to the surface safely, or to decompress the diver in a decompression chamber, is at least six hours.

As mentioned earlier, we can consciously decide not to breathe for a short period of time; this period can be significantly increased if the face and body are completely immersed in water. The increase is related to an innate physiological reaction called the **diving response**.

● Can you think of other mammals that might exhibit the diving response?

● Aquatic diving mammals, such as seals, dolphins and whales, can live under water for prolonged periods of time, sometimes at great depth, without the need to frequently surface for oxygen.

The diving response is readily apparent in infants and young children. As the child goes under water, they experience a physiological situation where there is no external O_2 available, i.e. total O_2 deprivation or *anoxia*. In this event, ventilation is stopped immediately and, within 30 seconds, stimulation of the vagus nerve to the heart initiates bradycardia (a slowing down of the heart rate; Section 2.4.6). The arterioles of all but the most vital organs constrict, ensuring that blood flow to the heart and the brain is maintained. Obviously, this situation cannot last indefinitely, but some children have been able to survive for up to ten minutes in cold water without drowning, and recover with normal brain function.

Finally, there are some humans who continue to emulate aquatic mammals by diving to seemingly unfathomable depths without oxygen and with only two lung volumes of atmospheric air (Phillips, 2001). The world record is over 160 m (i.e. over 525 ft!). At this depth, there is little daylight, it is very cold and complications due to the high external pressure frequently occur; these include: burst ear drums, compression of the lungs to about 6% of their volume at the surface, a massive shift of blood to the thoracic cavity due to the intense external pressure, impaired vision and heart rates of as low as four beats per minute. As you might imagine, unconsciousness is never far away. These sports divers are called (among several expressions) 'free-divers'. They deliberately expose the homeostatic mechanisms in their bodies to physiological and psychological terrains that are frighteningly hostile and dangerous in the extreme; this is ultimate sensation-seeking, with no room for a tickly cough, a sudden panic attack or 'a change of heart'. You may be curious as to how these people terminate their deep-sea immersion – they automatically inflate a balloon that hauls them up to the surface (and into the waiting arms of their colleagues). Alas, rules dictate that if competitors 'black-out' on reaching the surface, they are disqualified! What bad luck!

Summary of Section 3.5

1   The heart rate and respiratory rate can be altered to adjust to changes in the external environment and/or the state of the body.

2   Heart rate, stroke volume and thus cardiac output are increased during exercise to maintain blood flow to the active muscles. Ventilation and respiration are increased, so maintaining an increased supply of O_2 and lowering the increased levels of CO_2.

3   The body is able to acclimatize gradually to high altitudes, where the amount of O_2 in the air is reduced.

4   Decompression sickness is a problem for deep-sea divers when they return to the surface. The decrease in the pressure of water results in the formation of gas bubbles in the blood, which can block the nerve supplies to the brain.

5   When submerged, humans and aquatic mammals show a 'diving response', whereby heart rate slows and oxygen in the blood is only pumped to vital organs.

3.6 Conclusion

In this chapter, the main anatomical and functional features of the respiratory system have been presented. Of particular importance to understanding the operation of this system is knowledge of the cellular mechanisms of O_2 and CO_2 transport in the blood, and the internal and external factors that influence this transport, both in health and during abnormal conditions that affect lung function. The chapter considered specific lung diseases that seriously affect the healthy function of the respiratory system. Hopefully, the case reports presented in this chapter will have illustrated not only relevant clinical aspects, but also the far-reaching sociomedical implications of two important respiratory diseases.

We also hope that you will now have a better understanding of the integrative homeostatic control mechanisms that the respiratory and circulatory systems together exert on body function. Their concerted action was specifically illustrated in this chapter by examples indicating aspects of 'cardiorespiratory' adaptation to altered internal and external environments. The physiological fitness of the cardiovascular and respiratory systems are key factors limiting the successful physiological adaptation of the human body in occupying new environmental niches, both physiological and psychological. With this in mind, remember – aerobic exercise is good for your vitality…and this includes laughter!

Questions for Chapter 3

Question 3.1 (LOs 3.1 to 3.5)

Provide short descriptions of each of the following terms: (a) oxygen–haemoglobin dissociation curve; (b) the cause of pleurisy; (c) peak flow rate; (d) carboxyhaemoglobin; (e) Bohr shift; (f) sinus arrhythmia; (g) diving response.

Question 3.2 (LO 3.1)

Describe the pathway and mechanism by which air enters the body.

Question 3.3 (LO 3.2)

Give a short account of how O_2 and CO_2 are transported in the blood.

Question 3.4 (LO 3.3)

A friend experiences a sudden attack of asthma. Describe how anti-asthmatic drugs will alleviate their symptoms.

Question 3.5 (LOs 3.4 and 3.5)

What changes occur in the cardiovascular and respiratory systems in response to the onset of exercise and how are these changes brought about?

References and Further Reading

References

Hassed, C. (2001) How humour keeps you well. *Australian Family Physician*, **30**(1), 25–8.

National Asthma Campaign: 'Starting as we mean to go on.' An audit of children's asthma in the UK, *The Asthma Journal*, Special Supplement, May 2002, Vol. 8, Issue 2 [online]. Available at: http://www.asthma.org.uk/about/images/childrenaudit02.pdf (Accessed October 2004).

Phillips, H. (2001) Into the abyss, *New Scientist*, **2284**, 30–33.

Qian, W., Sabo, R., Ohm, M., Haight, J. S. and Fenton, R. S. (2001) Nasal nitric oxide and the nasal cycle, *Laryngoscope*, **111**(9), 1603–1607.

Sutoo, D. and Akiyama, K. (2003) Regulation of brain function by exercise, *Neurobiology of Disease*, **13**(1), 1–14.

Further Reading

British Thoracic Society [online]. Available at: http://www.brit-thoracic.org.uk/copd/pubs_sidenav.html (Accessed October 2004).

Cardiopulmonary Resuscitation (CPR) – The Harvard Medical School Family Health Guide [online]. Available at: http://www.health.harvard.edu/fhg/firstaid/CPR.shtml (Accessed October 2004).

CHAPTER 4

IMMUNOLOGY: DEFENDING LIFE

Learning Outcomes

After completing this chapter, you should be able to:

4.1 Briefly describe what is meant by 'innate' and 'adaptive' immunity.

4.2 Discuss the processes of phagocytosis and cytotoxicity.

4.3 Describe the functions and features of the inflammatory response.

4.4 Describe the processes of cell-mediated and antibody-mediated immunity (with particular reference to primary and secondary immune responses).

4.5 Explain what is meant by clonal expansion and clonal deletion.

4.6 Distinguish between artificially and naturally acquired immunity.

4.7 Distinguish between active and passive immunity.

4.1 Introduction: constant war

Our bodies are under constant attack, from the moment of our conception to our final days, from an enormous array of potentially harmful invaders. These invaders are diverse in nature and include such entities as bacteria, viruses, cancer cells (invaders from within), parasites and cells from transplanted tissue and organs. Our bodies possess a wide variety of protective strategies and these will be discussed in this chapter. These protective measures can be divided broadly into two *defensive* categories which collectively form our immune system. The first category includes the **innate immune system** (also referred to as the **non-specific defence mechanisms**) which can be thought of as the first line of defence that protects against a wide range of possible dangers. The second category is the **adaptive immune system** (also referred to as the **specific defence mechanisms**). Here the immune response is directed against only one particular invader and over time, this can lead to the formation of **immunological memory**, a process that confers long-term protection against a specific infection. Whilst it is convenient to view the immune system as two sets of defensive mechanisms, it is important to bear in mind that both systems work together and not in isolation.

Before we broaden our discussion in might be helpful to consider the consequences to us and to society when our immune systems fail or are under *potential* threat.

4.2 The Black Death

In the 14th century almost half of Europe's population was killed by the deadly bubonic plague. A fatal disease, caused by the bacterium *Yersinia pestis*, plague normally affects rats, but can be transmitted to humans by fleas. Once people were infected, they subsequently infected others very rapidly. Plague causes fever and a painful swelling of the **lymph glands** (also known as **lymph nodes**), small

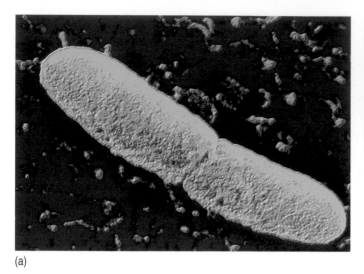

(a)

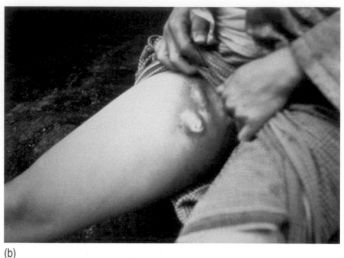

(b)

Figure 4.1   Bubonic plague (the 'Black Death'). (a) *Yersinia pestis*, the bacterium that causes bubonic plague. It enters the body via a flea bite or by inhalation of contaminated water droplets/sputum expelled by an infected human. (b) A hallmark of infection is the development of buboes, infected and greatly swollen lymph glands.

specialized organs that contain a high density of cells involved in conferring immunity (you came across these in Book 1, Section 1.6, Figure 1.10). These swollen lymph glands are called *buboes*, and this is how the disease gets its name (examples of *Y. pestis* and buboes are illustrated in Figure 4.1). The disease is also characterized by spots on the skin that are red at first and then turn black (hence the vernacular name, the 'Black Death').

The first major recorded outbreak of plague was in China in the 1330s, a period when China was at the centre of a vast trading network that reached as far as Europe and western Asia; it was only a matter of time before the plague spread. In 1347 the plague reached Europe, carried by a fleet of Italian merchant ships plying the waters between Sicily and the Black Sea, one of the key silk and porcelain trade routes between the Mediterranean and China. When the ships docked at the Sicilian port of Messina many of those aboard were already dying of the plague; within days the disease had penetrated the city and spread to the surrounding countryside. An eyewitness account from this time describes very graphically what happened:

> How many valiant men, how many fair ladies, breakfast with their kinfolk and the same night supped with their ancestors in the next world! The condition of the people was pitiable to behold. They sickened by the thousands daily, and died unattended and without help. Many died in the open street, others dying in their houses made it known by the stench of their rotting bodies. Consecrated churchyards did not suffice for the burial of the vast multitude of bodies, which were heaped by the hundreds in vast trenches, like goods in a ships hold and covered with a little earth.
>
> Boccaccio, *The Decameron*

By the following year the plague had reached England and nothing could be done to halt its relentless progress. As winter drew on the plague appeared to disappear, but only because the fleas that passed it from person to person became dormant during the cold winter months. Each spring, with the arrival of milder weather, the plague struck again, killing yet more people. After five years 25 million people were dead, approximately one-third of Europe's population.

Even when the worst was over, smaller outbreaks continued, not just for years, but for centuries. The survivors lived in constant fear of the plague's return, and the disease did not abate until the 1600s. The exact reasons for the end of the plague are not fully understood, though improvements in housing conditions, living standards and personal hygiene are all thought to have played a part. However, most researchers now believe that those who survived the plague did so because of genetic factors that improved their immunity to the disease. This is an example, then, of evolution by natural selection (Book 1, Section 1.4.2). If left untreated, 75% of those infected die and the remaining 25% recover and develop some degree of immunity. Today, plague is treated by a class of drugs called **antibiotics** – compounds that kill or disable invading bacteria.

Medieval society never recovered from the ravages of the plague. So many people had died that there were serious shortages of labour all over Europe. Workers demanded higher wages, a move which was initially resisted by the remnants of the pre-plague social order and by the end of the 14th century peasant revolts were widespread throughout Europe.

The influence of the Church was also challenged. Plague-struck Europe had prayed for deliverance, prayers that went largely unanswered. In short, the plague had ushered in a new period of political turmoil and philosophical questioning; a process that transformed the cultural, economic and social fabric of Europe.

Despite the passing of the Black Death, the peoples of Europe are still fearful of plague, but now in the context of germ warfare and bioterrorism. Indeed, some people believe that rogue nations and terrorist groups are stockpiling biological agents such as anthrax (a bacterial disease), small pox (a viral contagion) and bubonic plague.

4.3 The immune system – the combatants

Clearly, in the case of the Black Death we can think of *Y. pestis* as being the harmful invader and for some at least, the survivors of the plague, the immune system as being the defender. Figure 4.1a gives us an image of what an invader might look like, but what of the cells of the immune system? In this section we will introduce some terms used to define and describe harmful agents and invading microbes and also introduce the key cells that make up the immune system.

4.3.1 The invaders: pathogens and antigens

The human immune system has evolved the capacity to protect against a huge range of potentially damaging smaller organisms that may enter the body – some penetrating only the outer layers of skin, some reaching organs such as the brain, heart and liver, and others multiplying in numerous sites in the tissues and body fluids. A useful collective term for all of these potentially damaging smaller organisms is pathogen. They include many kinds of microbes – bacteria, viruses and fungi – together with single-celled and multicellular parasites, some of which may be very large (for example, tapeworms can reach over a metre in length). Perhaps one of the most notable parasites is *Plasmodium*, the microbe that causes malaria.

Any macromolecule, cell or organism that triggers an immune response against itself, if it gets into the body of a host animal, is known as an antigen (see Section 2.3.1 where antigen was discussed in relation to blood groups and rhesus factors). You should note that, although all pathogens are antigens and evoke an immune response in their hosts, many other kinds of antigens exist. Some of these are macromolecules secreted by pathogens (for example, the toxin produced by the tetanus bacterium), but others are macromolecules from non-pathogenic sources (such as plants and animals), or synthetic compounds made in the laboratory. All these molecular antigens are relatively large; equivalent to a smallish protein. Molecules smaller than this seem incapable of triggering an immune response unless they first bind to a protein in the body, thereby increasing their mass above the threshold. Cells are massively larger than this lowest size limit and are constructed from many different macromolecules, so cells can also be potent antigens.

However, the requirement to be larger than a minimum size is about the only thing that the huge range of different antigens have in common. **Antigenicity** (the ability to act as an antigen) is not an intrinsic property of the substance concerned, but depends on the ability of the *host* to recognize that substance as an antigen. Thus, antigen is a *relative* term: the same cell or macromolecule can behave as an antigen in one individual and not in another, and there are variations between individuals in terms of how 'strongly' they respond to the same antigen.

● From your experience can you think of a seasonal substance that acts as an antigen in some people and not in others?

● Grass pollen for some people is a powerful antigen – causing the conditions of hayfever and possibly asthma (Section 3.3.1).

Such allergies will be discussed more fully later.

The immune system can recognize each individual antigen *specifically* as a unique molecular entity, distinguishing one from another with exquisite precision. This property of the immune system is known as **antigen specificity**. Later in this chapter we will describe how antigen specificity is achieved by the non-specific and specific defence mechanisms, and point to similarities and differences in how these mechanisms recognize appropriate targets for attack.

The ability to recognize each antigen specifically also enables the immune system to achieve another vitally important function – the ability to *ignore* any cell or molecule that is *not* recognized as an antigen. This property is known as **immunological tolerance.** Thus the great majority of the human population possesses an immune system that maintains accurate recognition of certain substances as antigens throughout life, responding only to pathogens or other potential sources of harm, while displaying immunological tolerance to all other biological material.

● Why do you think it is so important that the immune system *only* responds to the genuine threats and is not triggered by harmless cells and molecules?

● Antigen specificity and tolerance are vitally important because the body of the host is itself composed of harmless cells and molecules; if these are mistaken for pathogens then the immune response would be directed towards damaging the host's own body. Another advantage of specificity is that many harmless particles derived from other organisms inevitably get into the human body – everything we eat, inhaled pollen grains, cat fur, house dust, etc. – which should ideally be 'ignored' by the immune system since they pose us no threat.

● Can you think of examples in which antigen specificity and tolerance breaks down, with damaging consequences for the host? (Book 1, Section 3.7.2 has an example.)

● In the so-called *autoimmune diseases*, such as rheumatoid arthritis, the immune system mistakenly recognizes normal tissues or molecules in the host's own body as antigens and attacks them. Similarly, type 1 insulin-dependent diabetes mellitus (you met this condition in Book 2, Section 3.7.2) is believed to be due to the autoimmune destruction of the insulin-releasing cells of the pancreas (the β cells). Hayfever is an example of another inappropriate immune response.

Tolerance to the cells and molecules of one's own body is known as **self-tolerance**. The ability of the immune system to detect and destroy hundreds of thousands of different kinds of pathogen, plus countless other antigens, while remaining self-tolerant, is one of its most fascinating as well as one of its most vital characteristics. Later in this chapter (Section 4.8), we will discuss the mechanisms that begin to take effect during the later stages of fetal development, which ensure that the immune system of the new-born baby is already self-tolerant when it emerges into a world crawling with pathogens.

As mentioned in the introduction, the immune system can be thought of as having two lines of defence, particularly in the way in which each system responds to antigen. The first, the innate immune system, also referred to collectively as the non-specific defence mechanisms (to aid understanding you might want to think of this as being the system which has *no memory* of the antigen; in other words, responds in a non-specific manner), reacts rapidly to the antigen, but is unable to 'memorize' it, so should the individual be exposed to it again the system has no memory of having met the antigen before. The second, the specific defence mechanisms or adaptive immune system, responds more slowly to infection but possesses a high degree of specificity; it is capable of *remembering* the antigen so that when it meets it again, it responds more potently and faster.

Whilst reading this chapter it is important to appreciate that the cells and molecules of both arms of the immune system interact. The adaptive immune system frequently incorporates cells and molecules of the innate system in its fight against harmful pathogens. For example, *complement* (a molecule used by the innate immune system – see Section 4.4.2) may be activated by *antibodies* (molecules of the adaptive immune system – see Section 4.7).

4.3.2 The defenders: the cells of the immune system

Throughout this chapter we will encounter a variety of cells that participate in protecting our bodies from attack. All of the cells in the immune system belong to a single 'family' called the leukocytes (from the Greek *leukos,* meaning 'white'), or colloquially the white cells. Leukocytes were traditionally called white *blood* cells because they were first detected in the bloodstream, but this term creates an incorrect impression: some populations of leukocytes are found in other tissues and many that do enter the bloodstream spend only a small proportion of their lifespan there.

Figure 4.2 shows that the origin of all mammalian leukocytes, erythrocytes and platelets is a single pool of precursor cells found principally in the bone marrow, called haematopoietic stem cells (Book 2, Section 3.3.1). These cells divide to give several different lineages of differentiated cells. As few as 30 of these stem cells are sufficient to regenerate the entire leukocyte and erythrocyte population of a mouse after the original cells have been destroyed by radiation; a similar approach is now being developed to achieve the same result in humans affected by radiation damage. You will need to refer back to Figure 4.2 many times as we proceed through this chapter.

Figure 4.2 The development of leukocytes (white cells) from haematopoietic stem cells in the bone marrow of mammals. There are three main lineages of leukocytes: the *lymphoid cells*, the *granulocytes* and the *monocytes*. Adaptive immunity relies on the small lymphocytes (in the lymphoid lineage), whereas innate immunity relies on all the other cell types. Resistance to infection requires complex interactions between all these groups of cells.

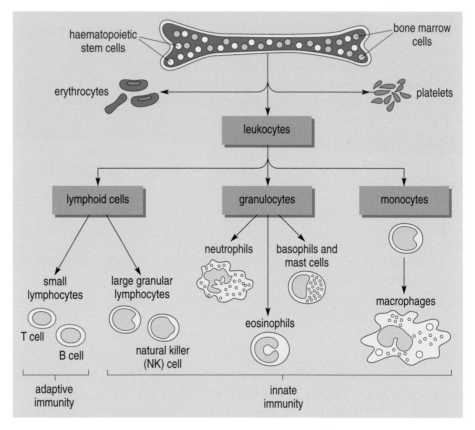

The names given to the different types of leukocyte generally tell you something about the function, biochemistry or morphology of the cell. For example, the lymphoid cells spend most of their lifespan in the lymphatic system, which consists of organs such as the lymph nodes, tonsils and spleen, connected by a network of lymphatic capillaries. We will describe the lymphatic system in more detail later because it is the major location of adaptive immune responses mediated by the leukocytes known as small lymphocytes. A great deal of the rest of this chapter will be about the small lymphocytes and adaptive immunity, but first we will concentrate on the leukocytes involved in the non-specific defence mechanisms.

Summary of Sections 4.1–4.3

1   The immune system has two branches: the first is innate, a built-in system of non-specific defence mechanisms; the second is adaptive and mounts a targeted and specific immune response (adaptive immunity is also called collectively the specific defence mechanisms).

2   Haematopoietic stem cells in the bone marrow divide and differentiate into leukocytes, erythrocytes and platelets.

3   Adaptive immunity relies on lymphoid cells known as small lymphocytes; all the other types of leukocyte contribute to innate immunity.

4.4 Non-specific defence mechanisms – innate immunity

The non-specific defence mechanisms form our first line of defence. They prevent the entry and limit the passage of a wide range of microbes and other material into the body. There are five main mechanisms and each will be discussed in turn. These mechanisms are often referred to collectively as *innate immunity*; they represent an inborn (innate) defence system that pre-exists the microbial invasion, offering a broad spectrum of defence against infection.

·   Defence at body surfaces

·   Constituent antimicrobial chemicals

·   Phagocytosis ⎤

·   Cytotoxicity ⎬ 'cell mediated'

·   Inflammation ⎦

The first two are independent of the immune system and the final three require the involvement of cells of the immune system.

4.4.1 Defence at body surfaces

Intact skin forms an impenetrable physical barrier to most pathogenic organisms and the acidity of sweat also inhibits the growth of many kinds of bacteria on the body surface. Some pathogens have evolved skin-penetrating delivery systems, as in the case of tropical parasites transmitted by mosquitoes and other biting organisms (e.g. the Black Death). It is important to appreciate that the skin is not the only surface that comes into direct contact with the outside world. The mucous membranes that line our respiratory and reproductive tracts and gut are also in direct contact with the external world. Here the mucus secreted by the mucous membranes traps microbes and other material on its sticky surface. In the lungs, the sweeping action of cilia moves mucus and inhaled particles towards the throat where it is cleared or swallowed. Hairs in the nose act as a coarse filter, reducing the amount of debris entering the lungs (Section 3.2.1). The one-way flow of urine from the bladder minimizes the risk of microbes ascending through the urethra into the bladder. In the mouth, the secretion of saliva washes away food debris that if left in the mouth would serve as a good source of nutrients for microbes. In addition, the slight acidity of saliva inhibits the growth of some microbes. Figure 4.3 (overleaf) summarizes the main physical and chemical barriers to invading microbes.

Figure 4.3 A summary diagram of the main physical and chemical barriers to infection in the human body. If these barriers are breached, then resistance depends on the competence of the immune system.

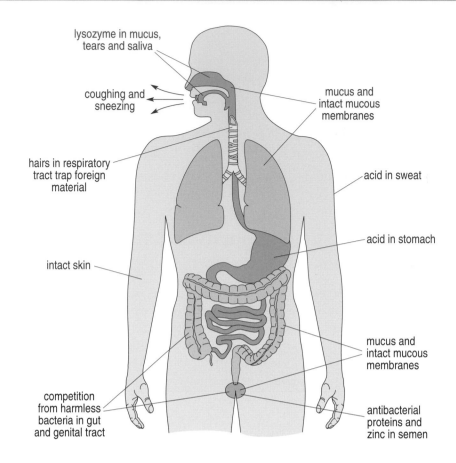

4.4.2 Constituent antimicrobial substances

The antimicrobial substances are a collection of molecules present in most body fluids and secretions that attack a wide spectrum of pathogens. Some are targeted towards bacteria whilst others engage viruses.

Lysozyme

An example of a *chemical barrier* to infection is an enzyme called lysozyme, which splits the chemical bonds that hold together the molecular components of bacterial cell walls. (Lysozyme was introduced in Book 1, Section 4.3.2.) Bacteria have a wall of polysaccharides outside their cell membrane. Some of the component sugar units of the wall are *unique* to bacteria and are the target for digestion by lysozyme, which damages the cell wall and thus kills the bacterium. Lysozyme is present in body fluids as diverse as blood, sweat, tears, nasal secretions, breast milk and the mucus lining the vagina and penis, and does no harm to human tissues because human cells do not contain bacterial sugar molecules.

Complement

A particularly important example of a chemical defence mechanism is a series of about 20 proteins collectively known as **complement**, which occur in certain body fluids including blood. The prime function of complement is to kill bacteria. (We will mention some of its other functions later.) The complement proteins are components of what is known as a *cascade reaction*. A cascade reaction relies on the presence of a series of molecules (or cells) that exist in an *inactive* state. They remain inactive until a specific trigger activates the first member of the series, which in turn activates the second member of the series, and so on, until the last member is activated. The final component of the series brings about a major change in the system. Generally, a scaling-up takes place as the sequence progresses, so that whereas only a few molecules or cells of the first member of the series are activated, they are able to activate many more molecules or cells of the second component, and so on. Figure 4.4 illustrates a typical complement cascade.

● What advantages can you see in cascade reactions?

◐ First, the reaction can very rapidly generate a high concentration of the final component, because all the elements are pre-formed in the system, i.e. they exist in an inactive precursor form, just waiting to be activated and scaling up occurs at each step. Second, the large number of steps in the sequence gives several opportunities for *negative feedback* to influence whether the cascade proceeds beyond that step. When the outcome of a completed cascade is a major change in the system, there is an advantage in having built-in fail-safe mechanisms that could switch it off at an early stage.

There are two main mechanisms for triggering the complement cascade. One of them is independent of the immune system, which is why it is included here as a chemical barrier to infection. Complement can be activated by the unique configurations of sugar units found only in bacterial cell wall polysaccharides; this is the *alternative pathway* of complement activation, so-named by early immunologists who saw this method as an alternative to the other, *classical pathway*, which involves antibodies and is described later.

Once activated, regardless of the method of activation, the proteins in the complement cascade build up on the surface of the target cell until many molecules of the final component are activated. This final component is a cylindrical assembly of proteins which form a tube, called the *membrane attack complex*. This tube punctures the surface membrane of the target cell, making a tiny hole. Complement usually attacks many different sites on the cell surface, so a large number of tiny holes can be punched through the membrane (see Figure 4.5).

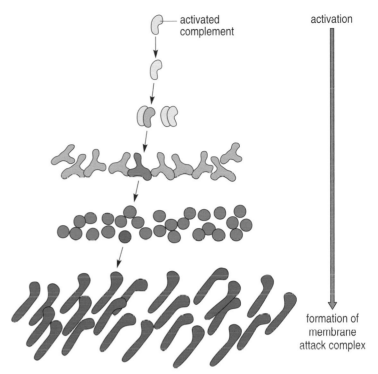

Figure 4.4 A stylized representation of a complement cascade. Note how activation of a single complement molecule initiates a series of events that leads to the amplification of the complement signalling system. The pattern of events is characteristic of a cascade mechanism.

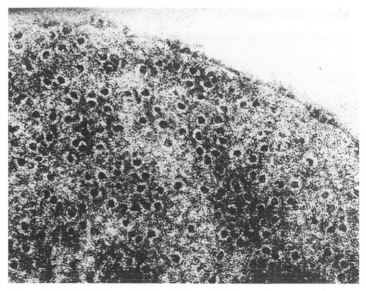

Figure 4.5 Holes punched through the surface membrane of a bacterium, *Escherichia coli* (*E. coli*), each of which is due to the membrane attack complex of activated complement.

- What do you predict will happen to a cell that has been punctured many times, as in Figure 4.5?

- The cell is unable to prevent fluids from outside the cell rushing in through the holes, so it swells and bursts. This process underlying the entry of water into the cell is osmosis, first mentioned in Book 1, Box 4.4.

Figure 4.5 dramatically illustrates why it is vitally important that the complement cascade is *only* triggered by potentially harmful cells such as bacteria, to avoid damaging the body's own cells.

Interferons

By contrast, the effectiveness of another important chemical defence, this time against viruses, rests on the ability to shut down the metabolic activity of normal cells in the body. When a cell becomes infected by a virus, it manufactures and secretes specialized proteins called **interferons**. Interferons are so named because they 'interfere' with the manufacture of new proteins. Viruses replicate by taking over the mechanisms of protein synthesis of the cell they have infected, diverting the cell to produce viral messenger RNA (mRNA) which is translated into viral proteins and assembled into new virus particles. Interferons damage viral mRNA and inhibit protein translation, not only in the infected cell, but also in neighbouring *uninfected* cells.

- How does the action of interferon defend the body against viral infections?

- It shuts down the ability to manufacture any proteins – host or viral – in cells neighbouring the original infection; this creates a barrier of cells which cannot be converted into 'factories' for manufacturing new virus particles and effectively walls up the infection inside this barrier.

Despite these physical and chemical barriers to infection, the sheer diversity and numbers of pathogens and their very rapid rate of reproduction means that additional defence mechanisms are essential to protect the integrity of the body. These additional mechanisms involve the specialized cells of the immune system, the leukocytes; you were introduced to these cells in Section 4.3.2.

4.4.3 Cell-mediated defence mechanisms: phagocytosis and cytotoxicity

Leukocytes are easily distinguished from other cells in terms of their shape and their susceptibility to various histological treatments. The *granulocytes* (the leukocyte lineage in the centre of Figure 4.2) have numerous dense structures in their cytosol which stain in characteristic ways when treated with different biological dyes used in the preparation of microscope slides. Granulocytes which do not stain with either acidic or alkaline dyes are known as **neutrophils**, whereas **eosinophils** stain with acidic dyes (of which *eosin* is one), and *basophils* and **mast cells** stain with alkaline (basic) dyes.

Neutrophils are the most numerous cell type in the bloodstream, where they contribute between 40% and 70% of the total 'white count'. The numbers rise sharply during an infection. The other granulocytes spend all or most of their lives in the tissues, particularly in areas close to the outside world such as the skin, respiratory system and gut. The dense granules from which granulocytes take their name contain a range of chemicals which inflict damage on pathogens, either directly, or indirectly by attracting or activating other components of the immune system. Neutrophils are also phagocytic (they are capable of digesting other cells – phagocytosis means 'cell eating'; you came across this term in Book 1, Section 2.10.1). They have an easily recognizable appearance, characterized by a multi-lobed nucleus (see Figure 4.6a). Neutrophils live only a few days and do not divide once they have entered the bloodstream.

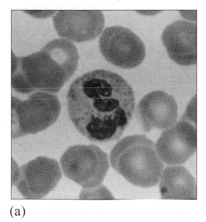 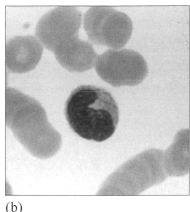

(a)   (b)

Figure 4.6   Stained leukocytes viewed through a light microscope. (a) A neutrophil with a characteristic multi-lobed nucleus; the pale cells in the background are red blood cells. (b) A macrophage with a characteristic kidney-shaped nucleus.

The *monocytes* (on the right in Figure 4.2) spend a relatively short time in the bloodstream before migrating through the capillary walls and into tissues and organs all around the body, where they differentiate into **macrophages**. Macrophage literally means 'big eater', from which you can deduce that these cells are phagocytes (in fact, the cell shown performing phagocytosis in Figure 2.23 of Book 1 is a macrophage). They are the largest of the leukocytes and have a characteristic horseshoe- or kidney-shaped nucleus (see Figure 4.6b). Less than 7% of leukocytes circulating in the bloodstream are macrophages, but they are found abundantly and widely distributed throughout the body tissues. Macrophages are highly mobile, long-lived cells that migrate through connective tissues by squeezing through the intercellular spaces. They are particularly abundant near blood vessels, the walls of the gut, in the genital tract, lungs, lymph nodes and spleen. Recall from Book 2, Section 1.9 that the microglia in the central nervous system perform a similar function to macrophages.

The *lymphoid cells* (on the left in Figure 4.2) are distinguished into two groups, one of which – the *large granular lymphocytes* – is involved in innate immunity. There are several different types of large granular lymphocyte, but we shall mention only one – the dramatically named **natural killer cell**, or **NK cell** for short. NK cells are particularly effective against virus-infected cells.

All of the cells involved in innate immunity can participate in an immune response without prior contact with an antigen. This property distinguishes innate from adaptive immunity. But, as you will learn later, the cells capable of innate immune responses are to some extent under the control of regulatory signals generated by the small lymphocytes (on the left in Figure 4.2).

The immune responses mediated by leukocytes can be thought of as having three kinds of defensive activity (though it should be remembered that they also work in unison with the specific defence mechanisms of the adaptive immune response):

(a) *phagocytosis*, i.e. engulfing potentially harmful material;

(b) *cytotoxicity*, i.e. killing infectious organisms by damaging the cell membrane of a pathogen or an infected host cell;

(c) generating *inflammation* around the site of an infection.

We will look at each of these in turn.

Phagocytosis

As mentioned earlier, *neutrophils* and *macrophages* are the main types of leukocyte primarily responsible for destroying pathogens by phagocytosis (eosinophils are also capable of phagocytosis but this is not their main function). The distribution of neutrophils and macrophages, predominantly in the blood and the tissues respectively, ensures that wherever infectious organisms enter the body they are likely to encounter a phagocytic cell. Phagocytosis by neutrophils and macrophages follows a similar course. The phagocytic cell must first adhere to the surface of the target, which may be a pathogen, cell debris, or an inert particle such as a splinter of wood.

Adherence may be enhanced by receptors on the surface of the phagocyte which bind to complementary shapes in the structure or surface of the target. Many bacteria have, as you already know, unique molecules in their cell walls which are not found elsewhere, and these act as binding sites for phagocytes. In addition, certain antibodies, components of the complement cascade and other specialized proteins in the bloodstream bind to many types of pathogen and label them as targets for phagocytosis. These chemical tags are known collectively as **opsonins**. Opsonins have two different kinds of binding site in their structure (see Figure 4.7): one site binds to the pathogen and the other binds to specialized receptors on the surface of the phagocytic cell. Thus, recognition and adherence may be effected by the same intermediary molecule (the opsonin) forming a stable bridge between target and phagocyte, which enables the phagocyte to get a good grip on the target. The labelling of pathogens or other targets with proteins that promote phagocytosis is known as *opsonization*, intriguingly derived from the Greek *opson* meaning 'cooked meat' or 'make tasty'. Throughout this course you will encounter the term 'receptor'. From what you have read so far you will have formed the view that a receptor in a cell membrane represents a specialized site that allows other molecules or cells to bind with it. Some receptors will bind a large variety of molecules whilst other receptors are highly specific and will only bind one type of molecule. Receptors perform many different tasks. Some provide a means of forming physical connections between cells, whilst others control physiological processes, such as the acetylcholine receptor at the neuromuscular junction (you came across these in Book 2, Section 4.8.3).

Adherence of the phagocytic cell to a target triggers a system of contractile filaments inside the phagocyte which enables it to throw 'arms' of plasma membrane (pseudopods; also shown in Book 1, Figure 2.23) around the target and enclose it in a vesicle (bag) of plasma membrane (see Figure 4.8). The vesicle containing the target is drawn into the cell where it fuses with numerous small *lysosomes*, intracellular packets of powerful degradative enzymes in an acid

Figure 4.7 Schematic diagram (not drawn to scale) showing that opsonins form a bridge between phagocytic cells and their targets, aiding the recognition and adherence that must precede phagocytosis and greatly increasing the rate of elimination of certain pathogens.

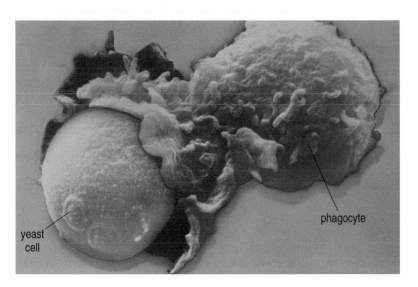

Figure 4.8 False-colour image showing a phagocyte engulfing a yeast cell. The yeast cell is ensnared by folds of plasma membrane, which draw it towards the phagocyte. The 'packaged' yeast cell is then taken into the phagocyte's interior, where it fuses with lysosomes and is broken down by powerful enzymes.

environment. These digest the target, ultimately into small molecules. (Beware! Lysosomes are not to be confused with *lysozyme*, the enzyme that digests bacterial cell walls; see Section 4.4.2.)

Cytotoxicity

Cytotoxic cells is a collective term for the many different cell types in the immune system that can kill other cells (such as pathogens) directly by chemical means, while lying alongside their target. Cytotoxic literally means 'cell poisoning' but the cytotoxic cells do more than this. Two examples from the innate immune system illustrate the general principles: these are the *eosinophils* and the *NK cells* (it may help you to look back at Figure 4.2), which use different methods to destroy their targets. (Note that there are also cytotoxic small lymphocytes which play a key role in adaptive immunity and will figure prominently in a later section of this chapter.) The cytotoxic cell must first adhere to its target and, just as in the case of the phagocytic cells, targets labelled with opsonins can be more effectively adhered to and are more susceptible to cytotoxic attack.

Eosinophils have numerous intracellular granules containing degradative enzymes and toxic chemicals which are directed out of the eosinophil onto the surface of the target cell. This may, in itself, be enough to damage the target cell membrane sufficiently to cause cell death. Some of these chemicals are also instrumental in generating an inflammatory reaction around the target, which will be described shortly. Eosinophils are particularly effective against the wide variety of parasites that live in the gut or burrow into the tissues.

By contrast, natural killer activity is directed against a very restricted range of targets, primarily the host's own body cells which have become infected with intracellular pathogens, particularly viruses. NK cells have receptors in their surface membrane that bind to viral glycoproteins; these appear on the surface of infected body cells as new virus particles are constructed inside them. NK activity can also be directed against certain tumour cells, but it is uncertain whether they are significant in controlling naturally occurring human cancers. The adherent NK cell releases several proteins into the small space between itself and its target, including a cylindrical molecule called *perforin,* which perforates the target cell membrane, opening up a pore to the outside.

- What is the likely result of the perforin's action?

- The result is the same as we described earlier when the final component of the complement cascade (membrane attack complex) punctures a bacterial cell, i.e. the cell swells and bursts.

4.4.4 Cell-mediated defence: the inflammatory response

When tissue is damaged it becomes inflamed and you will no doubt be able recall an incident when a cut or wound has become infected and inflamed (see also Book 1, Case Report 2.1). Inflammation is not an adverse effect, but represents a coordinated response to infection. Inflammation commonly occurs when the invading microbes have overcome the other non-specific mechanisms. Its purpose is to be protective; to isolate, inactivate and remove both the causative agent and the damaged tissue so that healing can take place. The classic external signs of inflammation are redness, swelling and heat (Book 1, Section 2.10). In the inflamed area, the walls of local blood vessels dilate and become very leaky, so that plasma and importantly, leukocytes flood out into the tissues; the area feels swollen and hot and looks red because of the increase in blood flow; the swelling can be painful and the area may tingle or itch. The flooding of the tissue also dilutes the toxic agents and assists in their removal from the site (particularly important when injurious chemicals and bacterial toxins are involved). Inflammation is a complex and coordinated sequence of events, with many different components. As it is usually short-lived, lasting a few hours or days, it is more correctly described as the **acute inflammatory response**, but often abbreviated simply to 'inflammation'.

- Can you suggest why an acute inflammatory response is, on balance, a good thing even though it is uncomfortable?

- Inflammation causes the area of an infection to be flooded with phagocytic and cytotoxic leukocytes and plasma, which contains a whole range of molecules that function in the immune response to infection (e.g. antibodies, lysozyme, complement). Despite the temporary discomfort, the inflammatory response can eradicate an infection before the pathogens have time to multiply enough to cause damage and disease. Even the pain may be a useful warning sign, prompting the person to protect the affected area.

As indicated earlier, inflammation occurs rapidly in an area of bacterial infection. This is because the inflammatory response can be initiated by the unique sugar units and glycoproteins found in many bacterial cell walls. These bacterial molecules cause activation of the components of complement, which, as mentioned earlier, are pre-formed in many body fluids. Some complement components are powerful promoters of inflammation; they attract leukocytes to the area (a process called *chemotaxis*) and trigger them into *degranulating*, releasing the contents of their granules which contain some chemicals with a wide range of effects on blood vessels and tissues. Inflammatory chemicals are an important component of innate immunity: some are toxic to pathogens; others attract, activate or immobilize leukocytes at the infection site; and some are opsonins, which facilitate adherence between phagocytic and cytotoxic cells and their targets.

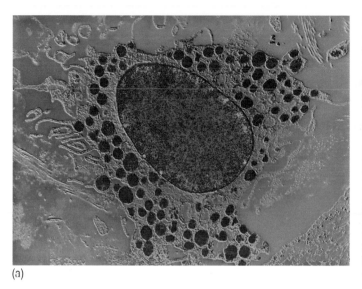
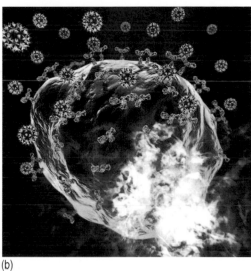

(a)   (b)

Figure 4.9   (a) A coloured electron micrograph of a mast cell. The cell has a large orange nucleus and green cytoplasm with numerous red granules which contain molecules involved in triggering and sustaining inflammation. (b) Computer artwork showing how, once activated, the mast cell expels the granules; despite the apparent disruption, it is not destroyed but recovers and synthesizes more granules.

All leukocytes are capable of contributing to inflammation, but the cells most often referred to in this context are the *mast cells* (see Figure 4.9). They are the most potent mediators of inflammation in the skin and in the membranes lining the respiratory tract and the eye sockets.

Unfortunately, the process of mast-cell degranulation is easily triggered in some people by harmless plant and animal material such as pollens and cat fur (Section 3.3.1). These *allergic* individuals make too much of a special kind of antibody (known as IgE and described in a later section), which binds simultaneously to the **allergen** (a general term for whatever substance the person is allergic to) and to mast cells, triggering the mast cells to expel their granules of irritant chemicals. One of these chemicals is histamine and consequently antihistamines are widely prescribed to treat allergies. Table 4.1 lists some of the principal substances released in inflammation.

Often acute inflammation, depending where it occurs, can cause considerable discomfort. The swelling associated with inflammation can have serious consequences. Swelling in a joint can severely limit movement and swelling in the larynx can affect breathing. Swelling in a confined region, such as under the skull where the increase in pressure is potentially life threatening (Book 2, Section 1.11.2), is often intensely painful. Pain associated with swelling occurs for many reasons; local swelling can compress the sensory nerve endings and this effect is then exacerbated by chemical mediators of the inflammatory process (see Table 4.1), for example bradykinin and prostaglandins enhance the sensitivity of the nerve endings to noxious stimuli.

An infected and inflamed wound often forms pus; a process called **suppuration**. Pus consists of dead phagocytes, dead cells, cell debris, fibrin (a protein involved in blood clotting), plasma and living and dead microbes. The most common bacterial microbes that cause inflammation and suppuration are *Staphylococcus aureus* and *Streptococcus pyogenes*. Small amounts of pus form boils and larger amounts

Table 4.1 Summary of the principal substances released in inflammation.

Substance	Source	Trigger for release	Main inflammatory response
Histamine	Mast cells (in most tissues), basophils (blood); stored in cytoplasmic granules	Binding of antibody to mast cells and basophils	Vasodilation, itching, increased vascular permeability, degranulation, smooth muscle contraction (e.g. bronchoconstriction)
Serotonin	Platelets Mast cells and basophils (stored in granules) Also in CNS (acts as neurotransmitter)	When platelets are activated and when mast cells/basophils degranulate	Vasoconstriction, increased vascular permeability
Prostaglandins (PGs)	Nearly all cells; not stored but made from cell membranes (as required)	Many different stimuli, e.g. drugs, toxins, other inflammatory mediators, hormones, trauma	Diverse, sometimes opposing, e.g. fever, pain, vasodilation or vasoconstriction, vascular permeability
Heparin	Liver, mast cells; basophils (stored in cytoplasmic granules)	Released when cells degranulate	Anticoagulant (prevents blood clotting), hence blood supply (nutrients, O_2) maintained to injured tissue and microbes and wastes can be washed away
Bradykinin	Tissues and blood	When blood clots, in trauma and inflammation	Pain Vasodilation

form abscesses. Both microbes release enzymes that affect the pattern of suppuration. *S. aureus* produces the enzyme *coagulase* which converts fibrinogen into fibrin, localizing the pus to the area of primary infection, but also impairing the ability of the leukocytes to reach the cause of infection. The strategy employed by *S. pyogenes* is quite different; instead these microbes release *streptolysin*, an enzyme that promotes the breakdown of connective tissue, allowing the infection to spread. Indeed, on rare occasions, if the spread of infection is left unchecked this can lead to a serious condition called **necrotizing fasciitis** – referred to in the media as the 'flesh eating bug disease!' Dan Gallon, a bus driver from St Louis in the USA, describes his encounter with *S. pyogenes* in Case Report 4.1 (Gallon, 2003). (You might also wish to re-read Section 2.10 on wound healing and Case Report 2.1 in Book 1 now.)

Infection is often associated with **fever**, an increase in body temperature, which in extreme cases can be fatal. The increase in body temperature is caused by the release of chemicals, called **pyrogens** (e.g. interleukin-1, a cytokine), from damaged tissue and the cells involved in the inflammatory response (particularly macrophages and granulocytes in response to microbial toxins and immune complexes). (Recall from Section 2.12 of Book 1 that this is not the only effect of interleukin-1.) Pyrogens are chemical mediators that act on the hypothalamus, which is involved in the homeostatic regulation of body temperature (Book 2, Section 1.4.2). The pyrogens activate a process that resets the hypothalamic thermostat to a higher temperature. The body responds by activating heat-promoting mechanisms, such as shivering and vasoconstriction, which enable body temperature to rise until the new level is reached. The increase in temperature increases the metabolic rate of the cells in the inflamed area and, consequently, there is an increased need for oxygen and nutrients. The increased temperature promotes the activity of phagocytes.

Case Report 4.1   Necrotizing fasciitis

On July 3, 2003, I had noticed one of my fingers had swollen and pus was oozing from it. It had swollen twice its size, so I put neosporin [antibiotic] ointment on it, covered it with gauze and went to work. I drive a bus for Metro, the transit company in St Louis, and was busy working overtime… My wife had cut my hair and as I was putting my shirt on, my infected finger brushed the back of my neck.

On July 6, the back of my neck and upper back had swollen. I woke up at 4 am that morning. My throat was closing, I was having trouble breathing, and could not wait for my wife to accompany me to the emergency room [Accident and Emergency Department]. I made it to the emergency room and was evaluated and subsequently admitted to the hospital. The doctor ordered pain medication and antibiotics to be administered through [intravenous] therapy. The Infectious Disease Team of doctors was called in…and changed the antibiotic I was receiving. The swelling continued and on Thursday July 10, I had an appointment at the Ears, Nose and Throat Clinic. The doctor looked at my neck and sent me back to my room. About 3:00 pm, the doctor came to my room and asked if he could aspirate some of the fluid from my neck. I agreed as long as it would alleviate the pain I was experiencing. Only about half a teaspoon of fluid was able to be aspirated and the doctor left my room. He returned about half an hour later with a consent form and requested my signature so he could lance my neck. I signed the consent form. When my neck was lanced, the pus drained uncontrollably. Six 4 × 4 gauze pads were used and then about 1 cup of fluid was also suctioned from the site. I might add the stench of the pus was unbelievable. A couple of hours passed – by that time it was 7 pm. The doctor returned with another consent form for surgery and said that surgery needed to be performed that night. The surgery team was on their way back into the hospital. I signed the consent for surgery and was relieved to know that I soon would not be in pain anymore.

That was only my first surgery. The next morning, Friday at 1 am, my doctor came and told my wife that they were checking into having me transferred to Barnes in St Louis in order to receive emergency medical care. Approximately half an hour later, the transfer was cancelled and another surgery was scheduled for 3 am. The second surgery was conducted by the surgery team – the first by the Ears, Nose and Throat team. It was at that time that the chief of surgery told my wife that I was a very very sick man. A member of the medical team told her I had a 50-50 chance of survival. (Little did I know my chance of survival was 20%.) Surgery was scheduled every day for debridement of the dead tissue and everyday I was in surgery at 8 am until five surgeries were completed to debride the dead tissue from the site. I was in [intensive care] for 3 weeks, on a ventilator, being fed through a tube down my nose and having a wound vac attached to my surgery site. The infection finally got into my bloodstream and at that point, my wife was informed that the antibiotic had to be changed to [one that] in all probability would shut my kidneys down. It was either change my antibiotic or die, so my wife said to change the antibiotic. The change worked, although the creatinine level in my kidneys went from 1 to 4.5 – 1 being normal. My kidneys had shut down – and this is part of sepsis – the gradual shut down of major organs. The nephrology team was called in and constantly was checking urine output as well as the creatinine levels. Gradually the creatinine level began to decrease.

I was released from the hospital on August 13, 2003 and am happy to be able to report that yes, I am a survivor of necrotizing fasciitis. There are many doctors and nurses that pulled my chart up on the computer just to follow my progress. And there were many doctors and nurses that would run into me and let me know how happy they were that I had survived.

My bout with necrotizing fasciitis was compounded by the fact that I have diabetes.

As mentioned earlier, large amounts of pus form abscesses. These can either be superficial or deep-seated. Superficial abscesses tend to rupture through the skin and void their contents (most of us at one time or another have helped this process along by squeezing the sides of an infected site!). The area affected by the abscess usually heals completely. Deep-seated abscesses on the other hand have a variety of outcomes and these are listed below.

- They can rupture completely, discharging their pus on to the surface of the skin, followed by healing.

- They rupture, but the discharge is limited and the infection develops into a chronic abscess with an infected open channel (*sinus*).

- They rupture and discharge into an adjacent organ or cavity, forming an infected channel that is open at both ends (*fistula*).

- The pus is eventually removed by the action of phagocytes, followed by healing.

- The pus-filled abscess can be enclosed by fibrous tissue, which in time becomes calcified. Whilst this isolates the infection, the microbes often survive within the enclosure and are a potential source for future infection.

Acute inflammation is usually short-lived, at worst no more than a few days. The damaged cells are cleared by the continued action of phagocytes. The waste material is removed in lymph (we will learn more about lymph in Section 3.3) and the blood and fibrin is broken down by enzymatic action. The wound generally heals, leaving very little scarring.

However, not all sites of acute inflammation are so readily resolved. Acute inflammation can develop into the chronic form if the initial infection is not fully cleared. This occurs when microbes remain active at the site of infection, particularly in deep-seated abscesses, wound infections and bone infections.

4.4.5 Chronic inflammation

The processes involved in chronic inflammation are very similar to those of the acute form but, because the process is of longer duration, more tissue is affected and destroyed. The leukocytes involved are mainly lymphocytes (we will learn about these in the next section) instead of neutrophils. In addition, fibroblasts are activated (Book 1, Section 2.9.4), resulting in the laying down of collagen, and fibrosis, the laying down of a protective fibrous layer that often leads to the formation of scar tissue. If the body's defences are unable to clear the infection, they may try to wall it off instead, forming nodules called granulomas, within which are collections of defensive cells. Tuberculosis is an example of an infection that frequently becomes chronic, leading to granuloma formation. The causative bacterium, *Mycobacterium tuberculosis*, is resistant to body defences and pockets of this microbe tend to become sealed up in granulomas within the lungs. A consequence of this is that tuberculosis can be difficult to treat, the walls of the granulomas forming a barrier against the delivery of antibiotics.

Summary of Section 4.4

1 Physical barriers such as intact skin and lining of the gut offer some protection against infection.

2 Chemical defences exist within the body quite independently of the immune system (e.g. saliva).

3 Lysozyme and the complement cascade attack the unique sugar units found only in bacterial cell walls.

4 Interferons suppress protein synthesis in cells around viral infections, so preventing the construction of new viruses.

5 The principal mechanisms employed by the cells of the innate immune system are: phagocytosis, in which pathogens and other targets are engulfed mainly by neutrophils and macrophages; cytotoxicity, in which the membrane of target cells is damaged by chemicals released mainly from eosinophils and NK cells; inflammation, caused by chemicals released mainly from mast cells and eosinophils and by components of the complement cascade.

6 Inflammation involves swelling, redness and heat as plasma and leukocytes leak out of blood vessels into an infection site.

7 Bacteria can release enzymes that affect the pattern of inflammation.

4.5 Specific defence mechanisms – adaptive immunity

4.5.1 Immunological memory

As mentioned earlier, the adaptive immune system is so named because it *adapts* whenever a particular pathogen (or indeed any antigen) is encountered for the first time; the response to second and subsequent invasions by the *same* pathogen is faster and more effective than on the first occasion – it reacts to that *specific* pathogen. In the early years of the 20th century, it appeared to immunologists almost as though the adaptive immune system was capable of 'remembering' all the pathogens it had encountered before, and changing its behaviour as a consequence of earlier experience. They called this property *immunological memory*, but we must emphasize that memory used in this context does not imply any similarity between the mechanisms of memory storage in the brain. Immunological memory is a property only of the adaptive immune system; the innate immune response to a particular pathogen is more or less the same no matter how many previous exposures to that pathogen the host organism has experienced.

4.5.2 The primary and secondary adaptive immune response

When a certain kind of pathogen gets into the body for the first time, a **primary adaptive response** (often abbreviated simply to *primary response*) occurs which is relatively slow to develop – it is barely detectable for about 7–10 days and then builds slowly to a peak within 2–3 weeks and lasts a few weeks at most

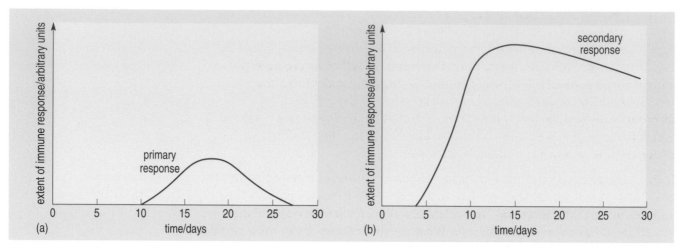

Figure 4.10 The general pattern of (a) a primary adaptive response to a particular pathogen, and (b) the secondary adaptive response to the same pathogen.

(Figure 4.10a). During this period, it is likely that the person will experience symptoms of infection that gradually get worse before they get better. Depending on the pathogen, medical intervention may be required. But if the *same* pathogen is introduced a second time, a greatly enhanced **secondary adaptive response** (or *secondary response* for short) occurs; it develops sooner, lasts longer and displays greater levels of activity than the primary response (see Figure 4.10b). The enhanced secondary response to a pathogen may be sufficiently effective to prevent any symptoms from developing, or the symptoms may be much milder than on the first occasion. If subsequent infections do not cause *any* symptoms, the person is said to be immune to that pathogen.

However, it is important to note that an individual who is immune to one kind of pathogen does *not* display heightened immunity to infection by other *unrelated* pathogens that have never been encountered before. Each type of pathogen is recognized by its characteristic 'marker' and responded to separately, in a specific way. The immunological memory relates only to those 'markers' that have been encountered before. These 'markers' are called *epitopes* and will be discussed in the next section.

Antigens, epitopes and antigen receptors

Earlier in this chapter we introduced the terms antigen and antigenicity. We will now refine our understanding further by discussing the molecular and chemical essence of an antigen – the *epitope*. First, it might be useful to refresh your memory:

● Can you recall the definition we gave earlier of an antigen?

● An antigen is any large molecule or cell that triggers an adaptive immune response against itself when it gets into the body of a host animal which, in turn, is capable of recognizing it as an antigen. Thus, the response of the *host* is central to determining the status of a cell or macromolecule as an antigen.

Variations in antigen recognition within and between species are determined partly by variations in genetic inheritance, which influence the ability of an animal to recognize and respond to material entering its body from the outside world, and partly by previous experience of exposure to antigens. You may recall that the adaptive immune system is so named because it adapts after an encounter with a particular antigen and becomes more effective at eradicating it on subsequent exposures. If we make the assumption that a certain cell or macromolecule *is* antigenic in a particular host, what features of the antigen does the host 'recognize'? And what distinguishes the antigen from the cells and macromolecules of the host's own body, or from harmless 'non-self' material to which the host should be tolerant? These questions are central to our understanding of how the immune system works.

It should be clear to you from our earlier discussion of target recognition that the small lymphocytes must have cell-surface receptors that bind to parts of the antigen, in much the same way as already described for leukocytes and their targets in an innate immune response. What is extraordinary about these **antigen receptors** is their *diversity*. Computer predictions based on an understanding of how they are constructed have produced an estimate that antigen receptors could come in as many as *100 million different shapes*, although each small lymphocyte can only make just *one* of these. The huge diversity of antigen receptors produced by small lymphocytes compares with the few *tens* of different receptor shapes that enable leukocytes in the innate immune system to recognize *their* targets (e.g. the sugar molecules found in the walls of bacteria).

It is clearly advantageous, in survival terms, for an organism to process a huge number of differently shaped receptors if there are a huge number of potentially life-threatening targets to which the receptors might have to bind. The cells and macromolecules of pathogens are biochemically complex, presenting millions of different permutations of smaller molecules on their surfaces. As you already know from our earlier discussion, every cell in the body and every invading pathogen is constructed from many different molecules and macromolecules, so cell surfaces are a mosaic of atoms and charges arranged in a dazzlingly complex array.

Imagine that this complex surface could be divided up into small areas, each composed of just a few adjacent molecules (e.g. about four or five amino acids, or sugars, or a few of each) lying closely together in a cluster. The number of *unique* biochemical clusters of this small size that could exist in nature on the surfaces of cells and macromolecules runs into *billions*. The task facing the organism is to distinguish between clusters which, despite any biochemical similarities in their structures, have vitally important differences in their meaning and significance. Those clusters that occur naturally in the organism's own body should be ignored by the immune system, even though many of these 'self' clusters *also* occur in the structures of pathogens. The immune system has no way of distinguishing between two identical clusters, one of which occurs in the host's own cells or macromolecules, while the other occurs in those of a potentially life-threatening pathogen. As a consequence, the immune system is *tolerant* to self clusters wherever they occur. Tolerance to clusters that are 'non-self' in the sense that they do not occur naturally in the host's own body is also advantageous for those clusters that occur only in harmless biological material such as pollens and foodstuffs. It is adaptive (in the evolutionary sense) for the organism to tolerate these intruders from the outside world, rather than commit resources to attacking them, since they pose no threat.

unique epitopes
on cell surface

Figure 4.11 Schematic diagram of the surface of a cell or macromolecule from a hypothetical antigen with several different epitopes, some of which overlap.

But survival also depends on the immune system being able to identify and attack all of the non-self clusters that occur *only* in the structures of pathogens or their harmful products (toxins), and which therefore act as unique signatures for these threats to life. These *antigenic* clusters of molecules are called **epitopes**. You can usefully think of epitopes as indicators that the whole of the cell or macromolecule on which they occur is both non-self and a legitimate target for an immune response. Antigenic cells and macromolecules inevitably have complex structures which usually generate more than one epitope, often several (see Figure 4.11). The same epitope can occur several times on an antigen and epitopes can even overlap – that is to say a molecule that contributes to one epitope can simultaneously contribute to a neighbouring epitope. The whole cell or macromolecule is referred to as an antigen, but much of the surface is not antigenic (i.e. does not trigger an immune response) because the biochemical clusters in these immunologically neutral areas also occur naturally within the structures of the host animal.

So, to recap, the adaptive immune system of humans (and most other vertebrate animals) is able to detect the presence of pathogens by recognizing and responding to epitopes on their surfaces, or in the structures of their harmful products (e.g. the toxins secreted by some bacteria). Moreover, the precision with which each epitope is identified as a unique entity means that the adaptive immune system can, in effect, distinguish between one kind of pathogen and another because the response is directed *only* against those epitopes that triggered the immune system into action. For example, the measles virus has certain unique epitopes in its structure which do not occur on other pathogens; an immune response against these epitopes is effectively an immune response directed specifically and exclusively against measles. This property is the basis of *antigen specificity* – one of the key terms we introduced in Section 4.3.1.

The act of recognition of an epitope by the adaptive immune system involves an interaction between the epitope and part of an *antigen receptor* with an exactly complementary shape, which protrudes from the surface of a small lymphocyte. This area is called the **antigen binding site**. (See Figure 4.12, but beware! The size of the receptors has been grossly exaggerated to make them visible; if this diagram was to scale, the small lymphocytes would have to be drawn as large as dinner plates before the microscopic antigen receptors could just be seen with the naked eye.)

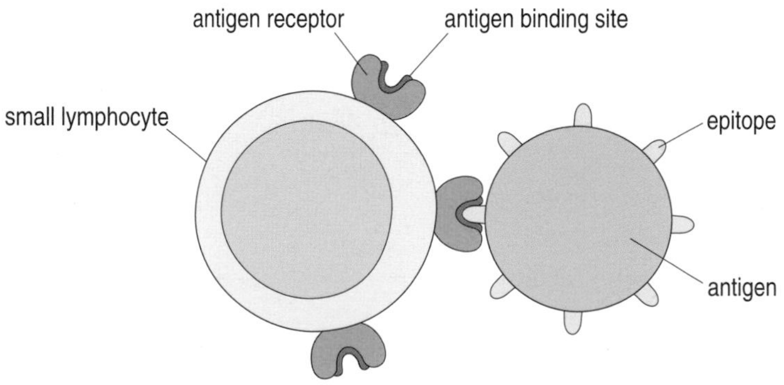

It has been estimated that at least 100 million *unique* epitopes exist on pathogens or their harmful products in the natural world. It follows that, in order to achieve a reasonable degree of protection against infection and parasitic invasion, a complex organism such as ourselves must be able to generate an equivalent diversity of antigen receptors, each with a different shape and charge profile in its binding site, each capable of binding to one of the epitopes in the structure of a pathogen or its products.

antigen receptor   antigen binding site

small lymphocyte

epitope

antigen

Figure 4.12 Schematic diagram showing a small lymphocyte with specialized antigen receptors on its surface (in reality there would be thousands of them). The antigen binding site exactly fits the complementary shape and charge profile of a particular epitope found on this hypothetical antigen, which could be a cell or a macromolecule.

In fact, each small lymphocyte can only manufacture a *single* shape of antigen receptor. If protection against pathogens is to be achieved, somewhere in the body there must be at least a few small lymphocytes with *pre-formed* antigen receptors of the correct shape to bind to one of the 100 million different epitopes. These few identical lymphocytes are said to be members of the same **clone**; the whole population of billions of small lymphocytes consists of roughly 100 million different clones. Between them they have sufficient antigen receptor diversity to bind to all of the epitopes on all of the pathogens we might encounter in a lifetime. In other words, our immune system is pre-programmed so that it is capable of reacting to every potential antigenic epitope.

● What would happen if we were attacked by a pathogen whose epitopes were not recognized as *self* or *non-self* by the adaptive immune system?

● If a clone did not exist that expressed the appropriate antigen receptor(s), then the adaptive immune system would not be able to mount an immune response (this would also be true if the pathogen only expressed '*self*' epitopes). Fortunately, the antigen receptors expressed by small lymphocytes are sufficiently diverse that such an eventuality is a near impossibility.

Another word of caution is in order at this point: you will often find the interaction between a small lymphocyte and an antigen (e.g. an infectious organism) being described as **antigen recognition**. This is an acceptable shorthand as long as you realize that the small lymphocyte is not 'recognizing' the antigen in the usual meaning of that term; the cell does not 'see' the antigen or 'decide' to go over to it and bind onto it.

● Try to write down a precise description of what antigen recognition really means in this example.

● We hope you wrote something like this. 'The small lymphocyte appears to recognize a specific antigen because binding is only possible between the binding sites of its antigen receptors and unique epitopes on the antigen which have a complementary shape. This enables the antigen receptors to bind to those epitopes and no others.'

4.5.3 Small lymphocytes and their territory

As mentioned earlier, the adaptive immune response centres on the activity of the leukocytes classified as small lymphocytes (a look back at Figure 4.2 may be useful at this point). They are recognizable under the light microscope because they have a very low ratio of cytosol to nucleus (see Figure 4.13), and what little cytosol there is has few organelles and no granules.

Small lymphocytes form about 35% of the leukocytes in the bloodstream, but they also circulate throughout

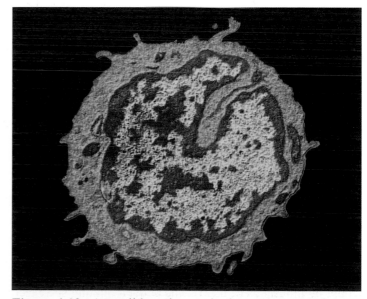

Figure 4.13 A small lymphocyte in the resting state, i.e. prior to contact with an antigen, viewed under an electron microscope. Note the sparse cytosol and organelles and the relatively large nucleus.

159

Figure 4.14 The human lymphatic system, showing the primary lymphatic organs (thymus, spleen and bone marrow) and the lymph nodes, connected by a system of lymphatic vessels and capillaries. The inset of the thumb illustrates the extent of the lymphatic capillary network.

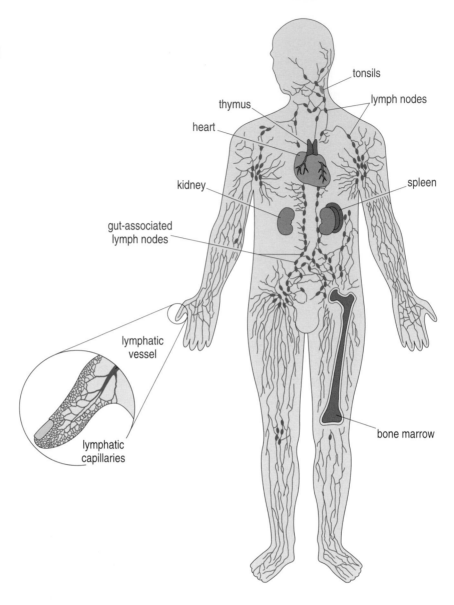

the lymphatic system and spend a considerable proportion of their lifespan stationary in the lymph nodes, spleen and other lymphatic organs shown in Figure 4.14 (see also Book 1, Section 1.6).

The lymphatic system is the main territory in which the adaptive immune response takes place. The fluid in the lymphatic capillaries is called lymph and it has a very similar composition to *plasma*, the fluid fraction of blood. The lymphatic system, which forms part of the circulatory system, has three important functions:

(i)   To defend the body against invading organisms by means of the immune system.

(ii) To collect and return interstitial fluid (the fluid between the cells) to the blood.

(iii) To absorb lipids from the digestive tract.

The lymphatic vessels are similar to veins of the cardiovascular system (see Section 2.3); the vessels have an endothelial lining supported by fibrous tissue and the larger vessels have muscle fibres in their walls. Interstitial fluid in the tissues passes through the walls of the lymph capillaries which are present in almost all capillary beds. The lymph capillaries join together to form the lymph veins or **lymphatics**. The lymph is propelled along the lymphatics by skeletal muscle activity and respiratory movements, as well as by inherent pulsating activity of the walls of the lymphatics. Valves prevent back-flow of lymph in a similar manner to the valves in the venous system. At intervals, the lymphatics empty into lymph nodes where the lymph is filtered and bacteria and other invading organisms are trapped and inactivated by cells of the immune system. The lymph leaves the nodes and returns to the cardiovascular system via ducts at the base of the neck which empty into the main veins draining the arms below the collar bone (the subclavian veins shown in Figure 2.9).

Interstitial fluid is formed as a result of a net loss of fluid from the capillaries as blood passes through the capillary network. The capillary walls are only one cell thick and so water and small molecules diffuse easily across them into the surrounding tissue. In addition to passive diffusion, the blood pressure within the capillaries forces fluid out through the wall. However, sizeable molecules such as large proteins, cannot pass through the capillary wall and are retained within the capillary forming what is called a colloid solution. Their presence creates a difference in osmotic pressure between the inside and the outside of the capillary and this is termed the *colloidal osmotic pressure* (look again at Box 4.4 in Book 1 for an explanation of osmosis). If the proteins were able to pass through the capillary wall, they would do so until the concentration of proteins on either side of the capillary wall was the same and an equilibrium established. However, since the capillary wall acts as a barrier to the proteins, they are trapped and there is a concentration difference between the inside and the outside of the capillary wall. Water is drawn into the capillary to balance this concentration difference and the force drawing the water in is the colloidal osmotic pressure (equivalent to about 25mmHg). The lymphatic system acts as a drainage system for about 10% of the interstitial fluid and any proteins that have leaked out from the capillaries.

- What would be the effect of a build-up of interstitial fluid in the tissue?

- The fluid balance in the body would be seriously disrupted, too much fluid would be present in the tissue and not enough in the circulating blood and the body would swell up like a sponge. [If the situation were to continue, death would occur within about 24 hours.]

Small lymphocytes recirculate between the vascular and lymphatic systems by the same route, but they can also actively burrow out through the walls of small veins as they pass through lymph nodes (Figure 4.15) and enter the lymphatic system directly. Note that they spend only a few *minutes* in the bloodstream during each circuit, compared with several *hours* in the lymphatic system.

Despite their homogeneous appearance under the microscope, there are several different types of small lymphocyte whose specialized functions in adaptive

Figure 4.15 (a) Scanning electron micrograph showing lymphocytes inside a capillary vessel passing through a lymph node. (b) Close-up of a lymphocyte starting its journey through the wall of the vein into the lymph node beyond. Lymphocytes are the only leukocyte that can leave the bloodstream and enter the lymphatic circulation by this route.

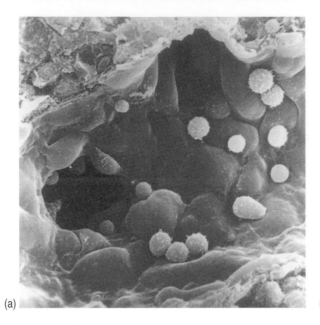

(a)

(b)

immunity are only revealed by their contact with *antigens*. We will describe these subsets of small lymphocytes later, once we have told you more about the antigens to which they respond, and about the remarkable cell-surface receptors that enable small lymphocytes to recognize antigens.

Diseases of the lymphatic system

Cancerous diseases which start in the lymphatic system are called *lymphomas*. One group of lymphomas was first described by Thomas Hodgkin in 1832 and is known as Hodgkin's disease (HD). A defining feature of Hodgkin's lymphoma is the presence of a distinctive abnormal lymphocyte called a *Reed–Sternberg cell*. Hodgkin's lymphoma is normally treated by radio- or chemotherapy and has a very high cure rate, especially in younger patients diagnosed with 'early-stage disease'. In this group the cure rate may approach 100%. The other group, which do not have the features of HD, are known as non-Hodgkin's lymphomas (N-HL). Patients with N-HL are treated with a combination of anti-tumour drugs and most survive following early diagnosis.

4.5.4 B and T lymphocytes

We mentioned earlier that different types of small lymphocyte exist, each performing a specialized function in adaptive immunity. Despite their similar appearance when 'resting', major differences in their activity are revealed as soon as they are 'activated' by the binding of complementary epitopes to their antigen receptors.

There are two main families of small lymphocytes, both derived by cell division and differentiation from the haematopoietic stem cells – the originators of all leukocytes, erythrocytes and platelets. These families are called the B and T lymphocytes respectively (look again at Figure 4.2). The **B lymphocytes** (or simply **B cells**) mature in the bone marrow (B is for bone marrow). Their role in the immune response is to produce antibodies (see Section 4.6.1), proteins designed to bind to, and cause the destruction of an antigen. The antibody released reacts with only one type of antigen and no other (hence the specific nature of this defence mechanism). The B lymphocytes provide *antibody-mediated immunity*.

The **T lymphocytes** (or **T cells**) mature in the thymus (T is for thymus, a lymphatic organ lying just above the heart, see Figure 4.14). The thymus releases a hormone, *thymosin*, which promotes the formation of fully differentiated and functional T lymphocytes. T cells have diverse roles. Some have a regulatory role that affects all aspects of immune responses (both adaptive and innate); some are involved in sustaining prolonged inflammation and others kill any of the host's own body cells that become infected. A mature T lymphocyte, like the B lymphocytes, has been programmed so that it can only recognize one type of antigen, so during its subsequent travels through the body it will react to no other antigen, no matter how dangerous it may be. Thus, a T lymphocyte manufactured to recognize the chickenpox virus will not react to the common cold virus, a cancer cell or *Y. pestis*. T lymphocytes provide *cell-mediated immunity*.

From the brief description of the T and B lymphocytes given above, it is clear that for every one of the millions of possible antigens that might be encountered in life there is one corresponding T and B lymphocyte. There is therefore a vast number of different T and B lymphocytes in the body, each capable of responding to only one antigen – or more specifically, to a single epitope. If you recall there are millions of different types of lymphocyte clones, each corresponding to a specific *epitope*. The process of antigen recognition at the level of a single clone (a process called *clonal selection*) ensures that the immune system mounts a highly specific and appropriate response to the invading pathogen.

Clonal expansion

The few small lymphocytes in each clone which are present at birth are not sufficient to mount an effective immune response against an infection, which normally requires millions of lymphocytes. The ability to scale up the number of cells committed to an immune response is called **clonal expansion** – a process which also underlies *immunological memory*.

When an antigen enters the body and the matching clones of small lymphocytes bind to it, activating signals are produced and these few cells divide; their daughters divide in turn, and the process is repeated many times over so that a rapid clonal expansion occurs (see the upper panel of Figure 4.16). The expansion in cell numbers is remarkable: in just four days, a single activated lymphocyte could, in theory, give rise to roughly 64 000 daughter cells, all of which are members of the same clone and have antigen receptors for the same epitope. The cells also mature and *differentiate* (become specialized for particular tasks) during this expansion phase. Most of the new cells formed are immunologically active cells with a variety of defensive functions, which eliminate the antigen in ways that we will describe later.

When the defensive cells have destroyed the antigen, most of them die, so each clone of small lymphocytes first expands and then contracts as the need for an immune response wanes. But some of the new cells formed during the primary immune response differentiate into so-called **memory cells**. The memory cells

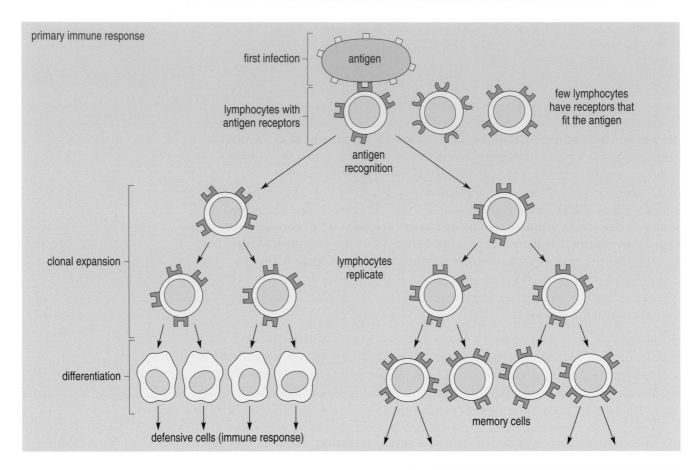

primary immune response

first infection — antigen

lymphocytes with antigen receptors

few lymphocytes have receptors that fit the antigen

antigen recognition

clonal expansion

lymphocytes replicate

differentiation

defensive cells (immune response)

memory cells

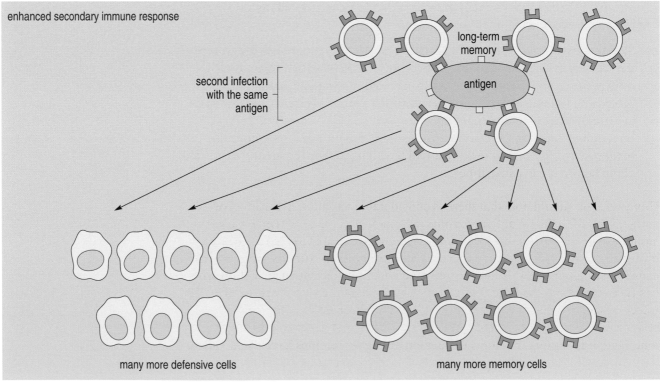

enhanced secondary immune response

second infection with the same antigen

long-term memory

antigen

many more defensive cells

many more memory cells

◀ **Figure 4.16** Clonal selection, clonal expansion and differentiation during a primary and a secondary immune response. The defensive cells participate in the immune response against an antigen (here shown with just a single epitope shape), and the long-lived memory cells ensure a more rapid and effective secondary response.

have a long life and continue to circulate in the body for months after the original exposure to the antigen is over and the defensive cells have died. They carry the same antigen receptors as the original members of the clone, so they permanently enlarge the clone of small lymphocytes capable of recognizing the epitope that triggered the immune response in the first place. This enlarged clone is able to respond much more rapidly by clonal expansion if the same epitope is encountered again (see the lower panel of Figure 4.16). This ability underlies the greater effectiveness of the *secondary* immune response (Figure 4.10) and explains how immunological memory of previous encounters with antigen are stored – not in the brain, but in expanded clones of cells with antigen receptors capable of binding to previously encountered epitopes.

Each time the same antigen enters the body, the corresponding clones are reactivated and become even larger. Even without subsequent contact with the antigen, memory cells generally replicate at intervals to replenish their numbers, so that, in effect, they may offer life-long protection against a certain epitope, and hence against any antigen which has that epitope in its structure.

● When children are immunized against common infectious diseases, such as whooping cough and diphtheria, they are usually given three injections of the vaccine at intervals of several months. Why do three, spaced injections give better protection than one?

● The vaccine contains samples of the unique epitopes found in the structure of the pathogens you wish to protect the child against. The first injection triggers off a primary immune response, in which the few lymphocytes with antigen receptors capable of binding to those epitopes multiply (clonal expansion) into an enlarged clone containing long-lived memory cells. The second and third injections trigger the production of even more memory cells, so that if the live pathogen ever gets into the body it will be eliminated by the greatly expanded clone before it can cause illness.

However, you should note that there is considerable variation in the level of antigenicity of different antigens: some are much better than others at triggering clonal expansion and memory cell formation and hence at evoking long-lasting protection against subsequent exposures. For example, the rubella virus, which causes German measles, is highly antigenic and, as a consequence, a child who has recovered from the illness rarely contracts it again. By contrast, most of the infectious agents that cause sexually transmitted diseases (e.g. the bacteria responsible for gonorrhoea and syphilis) are weakly antigenic and do not evoke long-lasting immunity; attempts to develop effective vaccines have so far failed.

Summary of Section 4.5

1 The lymphatic system is part of the circulatory system and consists of a network of vessels, lymph glands and lymphatic fluid (lymph).

2 The lymphatic system returns excess fluid to the blood, transports lipids from the gut, and filters and destroys microbes and other particles in the body.

3 Excess interstitial fluid which has passed out of the capillaries and is not absorbed by the tissues is collected by the lymphatic capillaries and returned to the blood.

4 Cancers of the lymphatic system are called lymphomas – e.g. Hodgkin's disease.

5 Antigens have small areas in their structures known as epitopes, which do not occur naturally in the body of the host animal, and against which it is capable of making an immune response.

6 Small lymphocytes circulate around the vascular (blood) and lymphatic systems, carrying antigen receptors on their surfaces with binding sites that fit epitopes.

7 Each small lymphocyte belongs to a clone of identical cells; each clone has receptors for only one of perhaps 100 million different epitopes.

8 On the first encounter with an antigen, the few small lymphocytes with complementary antigen receptors bind to its epitopes (clonal selection).

9 During the primary adaptive response, the selected lymphocytes undergo clonal expansion and differentiate into *defensive cells*, which eliminate the antigen and then die, and *memory cells*, which survive to form an expanded clone.

10 This clone responds more effectively during a secondary response to the same antigen, and is the basis of immunological memory and the protective effects of immunization.

4.6 T lymphocytes and cell-mediated immunity

The T cell population is divided into two major *subsets* (i.e. cytotoxic and helper), each with specialized functions (see Figure 4.17).

The **cytotoxic T cells** use toxic and perforating chemicals to kill the body's own cells when the latter have become infected with intracellular pathogens (those that multiply inside a host cell, out of reach of antibodies and innate immunity), in much the same way as described earlier for NK cells. The **helper T cells** synthesize and secrete a range of signalling molecules with activating or enhancing effects on most other cells involved in both the adaptive and innate immune responses. The memory cells provide cell-mediated immunity by responding rapidly to another encounter with the same pathogen (the role of memory cells in immunity was mentioned earlier).

The signalling molecules produced by helper T cells belong to a structurally and functionally diverse group of molecules, collectively known as *cytokines* (you were introduced to cytokines in Book 2, Section 3.8). The cytokines produced by

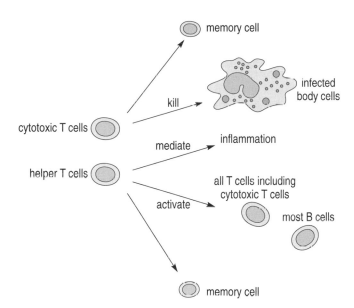

Figure 4.17 The two main T cell subsets and their general functions. Note that these subsets are further divided into groups of cells with specific roles but detailed explanation of all of these is beyond the scope of this chapter.

T cells are immensely important, but it is worth noting that macrophages also synthesize cytokines that are essential for T cell activation – another example of the interdependency of the innate and adaptive immune systems. These cytokines are all peptides (short chains of amino acids). They have a very short range, operating over the minute distances between cells in close contact, i.e. acting in a paracrine fashion (Book 2, Section 3.8); and they have a very short life – they survive only a few minutes in some cases before being broken down into their constituent amino acids.

Figure 4.18 illustrates how extensively the mechanisms of innate immunity and the production of antibodies depend on activating signals transmitted by cytokines from the helper T cells. The reliance of all these defensive mechanisms on helper T cells and their cytokines underlies the devastating effect of the human immunodeficiency virus (HIV) which destroys helper T cells, while leaving the rest of the immune system intact. In the absence of helper T-cell cytokines, none of the other mechanisms work effectively and the affected person gradually succumbs to a range of other infections.

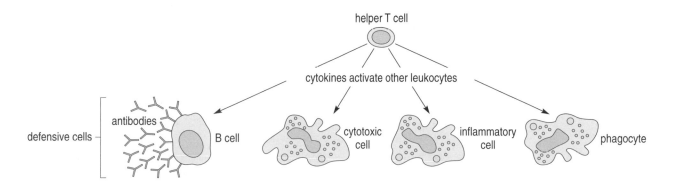

Figure 4.18 Both the innate and adaptive mechanisms in the immune system are regulated by the activity of the helper T cells. (Other subsidiary regulatory mechanisms exist, but these cannot compensate for the loss of normal T cell function.)

4.6.1 The T-cell antigen receptor

It should be evident by now that all T cells – whatever their subset – must carry antigen receptors anchored in their surface membrane with binding sites for particular epitopes facing outwards. If the antigen has come from outside the body, it needs to be 'presented' to the T lymphocyte on the surface of another cell (the antigen-presenting cell). The T-cell antigen receptor has unique requirements in the manner in which the epitope has to be presented to the T cell before binding between the two can take place.

The central structure in a T-cell antigen receptor consists of two short peptide chains (polypeptides), which together contain a binding site for a particular epitope. Close beside them are other essential polypeptides, which *restrict* the T cell from binding to an epitope except when it is being presented to the T cell by one of the *host's own cells*. This aspect of antigen recognition by T cells is one of the most difficult concepts to grasp for students new to immunology, but it is of fundamental importance.

Several different types of leukocytes, including B cells and all the phagocytic cells (e.g. macrophages), can process antigens into a form that the helper T cells can respond to. Collectively, all the cells that can perform this function (in addition to their other duties) are known as **antigen-presenting cells**, or **APCs** for short. They bind to the antigen via their own receptors and draw it into the cell; then using the enzymes in their lysosomes, they process it by breaking it up into fragments, just as we described earlier as the outcome of phagocytosis (Figure 4.19). But the next step may surprise you: antigen fragments are transported back to the cell surface and displayed there. Some fragments contain a single epitope, which is displayed in the cleft of a much larger molecule known as an **MHC molecule**, rather like a hot dog sausage being held out in a bun! (We will explain later what the letters MHC stand for; it would be a distraction here.) Helper T cells can only bind to epitopes which have been processed and presented to them in this way.

The processing of antigens for presentation to cytotoxic T cells is somewhat simpler. Cytotoxic T cells are adapted to kill the host's own cells which have become infected by intracellular pathogens. The epitopes of intracellular

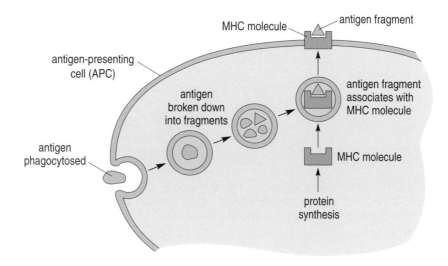

Figure 4.19 Schematic diagram showing antigen processing by an antigen-presenting cell (APC). Fragments containing an epitope are displayed on the cell surface in the cleft of an MHC molecule.

pathogens reach the surface of the infected cell when pathogens die and bits of their structures are expelled through the cell membrane; some of these fragments end up in the clefts of MHC molecules on the infected cell's surface. Additionally, viruses reproduce by assembling their component molecules at the surface of the cell they are infecting and, inevitably, some of these components find their way into the MHC clefts. In this location on the surface of an infected cell, the cytotoxic T cells are able to bind to the epitopes presented to them.

You may be wondering why this complex arrangement of presenting T cells with epitopes nestling in MHC molecules has evolved; a clue lies in Figure 4.17, which we suggest you look at again.

● Consider the functions of the main T cell subsets (cytotoxic and helper) shown in Figure 4.17. What do they have in common?

● They all make contact with other 'self' cells: the cytotoxic T cells kill infected body cells, while the helper T cells interact with most other kinds of leukocytes in the immune system.

It is therefore essential that T cells accurately recognize which of the cells they encounter is 'self'. The MHC molecules can be thought of as markers of 'selfness'. Each person has a unique configuration of MHC molecules on his or her cells, so a T cell can recognize any cell it encounters as self, i.e. from the same individual as the T cell. (Identical, i.e. monozygotic, twins are the only exception to the rule that each individual's MHC molecules are unique; identical twins have the same MHC molecules, so the T cells of one twin would recognize the other's cells as self.)

It would be energetically wasteful if T cells were activated every time they bumped into another self cell. T cells have evolved a sort of dual key activation mechanism which ensures that they are only activated if they encounter self-plus-epitope. In consequence, the T-cell antigen receptor can only bind to an epitope when it has been removed from the whole antigen and is then presented to the T cell in the cleft of an MHC molecule (Figure 4.20). T cells are referred to as **self-MHC restricted** in their activity.

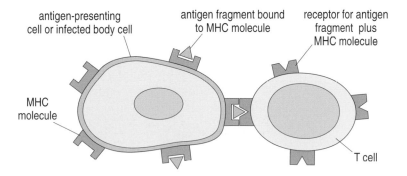

Figure 4.20 Schematic diagram of a T cell binding to an epitope which is being presented to it by a self cell in the cleft of an MHC molecule. Part of the T-cell antigen receptor binds to the epitope and part binds to the MHC molecule. The self cell could be either an antigen-presenting cell, or alternatively a cell which has been infected by pathogens, whose antigen fragments appear on the surface.

● What advantage can you see in the evolution of *helper* T cells which are restricted to recognizing self-MHC plus epitope?

● Self-MHC restriction limits each helper T cell to interacting only with those leukocytes in the immune system that have *already* bound to the epitope for which the T cell also has antigen receptors. It prevents the helper signals from being wasted on leukocytes that have yet to meet their antigen, or on cells that have no part in an immune response.

● What advantage can you see in the evolution of *cytotoxic* T cells which are restricted to recognizing self-MHC plus epitope?

● Self-MHC restriction ensures that cytotoxic T cells can only kill self cells that are harbouring intracellular pathogens. This property also enables them to identify an important target that might otherwise be missed.

The value of self-MHC restriction is well illustrated by the interaction between B cells and helper T cells (Figure 4.21). First the B cell binds to the intact antigen via the binding sites in its molecules of surface antibodies (we will learn more about these in the next section); then the antigen is internalized by the B cell and processed into fragments before reappearing at the surface in association with MHC molecules. Epitopes presented in this way to the matching clone of helper T cells (i.e. those with antigen receptors with binding sites for epitopes on the *same* antigen), together with activating signals (cytokines), cause the immediate activation of the helper T cell. In return, it sends activating signals back to the B cell which enable it to begin cell division and differentiation to produce a clone of plasma cells (we will meet these in the next section), which synthesize and secrete large amounts of antibodies – all of which also have binding sites for the same epitope.

Finally, you may be interested to learn a little more about the MHC molecules. MHC stands for **major histocompatibility complex**. Although the name is hard to grasp at first sight, the middle word – histocompatibility – gives a clue about how these molecules were discovered: *histo* means tissue, hence tissue-compatibility. MHC molecules are markers of self and, if tissue is transplanted between non-identical individuals, the recipient's immune system recognizes the grafted cells as non-self because their MHC molecules differ from those of the recipient. If the MHC molecules of donor and recipient are completely different, the two individuals are said to be histo-incompatible, and the graft will be attacked and rapidly destroyed by the recipient's immune system. If there is some similarity between the MHC molecules of the two individuals, then (with help from drugs that inhibit immune responses) the graft may survive. Thus, MHC molecules hold the key to the rejection by the immune system or the survival of tissue grafts or organ transplants between non-identical individuals. The MHC molecules were discovered by immunologists investigating the mechanisms underlying graft and organ rejection in the early days of transplant surgery.

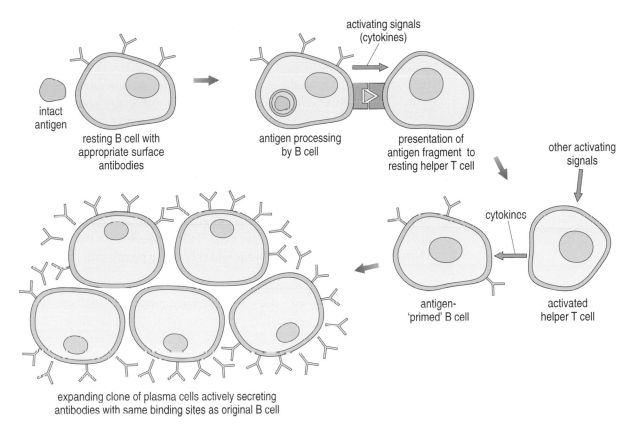

Figure 4.21 Summary diagram of the interaction between a B cell and T cell, both of which have antigen receptors for epitopes on the same antigen. The B cell binds to, internalizes, processes and presents the antigen to the helper T cell, and transmits activating signals to it. Activating signals (cytokines) from the T cell initiate clonal expansion of the B cell, culminating in plasma cell differentiation and antibody production. All of the cells involved and all of the antibody molecules have binding sites for epitopes on the *original* antigen.

Summary of Section 4.6

1   In order for T cells to be activated by antigen, the antigen must first be processed and presented by an antigen-presenting cell (APC).

2   The processed antigens are presented in the cleft of the MHC on the surface of APCs.

3   Cytotoxic T cells bind to epitopes presented in the MHC molecules on the surface of self cells, which are infected with pathogens.

4   Self-MHC restriction ensures that the helper T cells interact only with other leukocytes which have already encountered their matching antigens; they will only kill infected self cells.

5   Cytokines released from helper T cells regulate innate and adaptive immune responses.

4.7 B lymphocytes and antibody-mediated immunity

B lymphocytes, unlike the T lymphocytes which are free to circulate around the body, are fixed in lymphatic tissue (e.g. the spleen and lymph nodes). B lympho-cytes, unlike T lymphocytes, recognize and bind antigenic epitopes without having to be presented to them by an antigen-presenting cell. After contact with antigen they undergo clonal expansion and differentiate either into memory B cells, or into a type of defensive cell called **plasma cells**, which secrete large quantities of the specialized proteins known as antibodies. You should bear in mind that millions of unique clones of B lymphocytes and T lymphocytes exist in the body.

The antigen receptors on all B cells are known as *surface immunoglobulins*, denoting that they are globular proteins with an immunological function and that they are carried on the surface membrane of the cell. B cells are the only type of cell in the body to have surface immunoglobulins; each cell has about 100 000 copies of the *same* molecule on its surface. It is convenient to think of these molecules as Y-shaped, with a 'tail' region attached to the B cell and two 'arms' facing outwards. B cells bind to an antigen via either or both of two antigen binding sites in each surface immunoglobulin molecule, one located at the tip of each arm; these binding sites interact with complementary epitopes on the antigen (see Figure 4.22a).

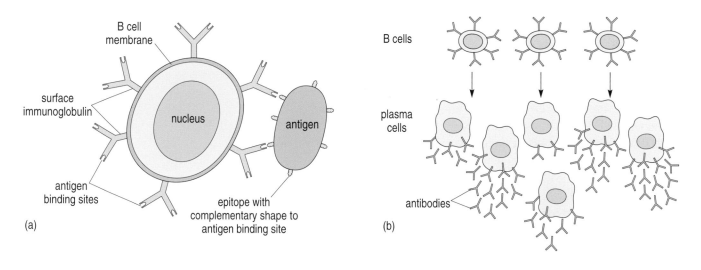

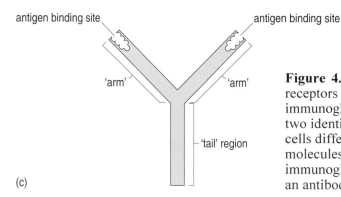

Figure 4.22 Schematic diagrams (not to scale). (a) The B cell's receptors for antigen are molecules of a protein known as surface immunoglobulin embedded in the cell membrane; each molecule has two identical antigen binding sites facing outwards. (b) Activated B cells differentiate into plasma cells, which secrete immunoglobulin molecules with the same binding sites as the original B cell's surface immunoglobulins. (c) A secreted immunoglobulin molecule is called an antibody.

As mentioned earlier, antigen binding, together with certain activating signals from other cells in the immune system, particularly the helper T cells, causes the original B cell to divide and differentiate into an expanded clone of memory B cells and plasma cells. Plasma cells synthesize and secrete large amounts of immunoglobulin molecules which are almost identical to the surface immunoglobulin molecules that the original B cell used as its surface receptor for antigen (Figure 4.22b). These secreted immunoglobulins are the *antibodies* that appear in the bloodstream and tissue fluids.

One final point about the B-cell antigen receptors (i.e. surface immunoglobulins) should be emphasized. These receptors can bind to epitopes only when they occur on what is known as *native antigens*, a term used to signify that the antigen is *intact*, for example, a pathogen or harmful macromolecule as it occurs in life. Thus, B cells recognize and respond to native antigens; this is very different from the *processed antigens* that T cells bind to, which have to be partly broken down and presented to the T cell in a specialized way before binding can take place.

Antigen binding is an *essential* requirement for B cell activation and for a minority of antigens it is also a *sufficient* activating signal. However, most B cell clones require antigen binding *plus* additional signals from *T cells* before they can begin clonal expansion. Before moving on, consider for a moment how many complex cellular processes are subsumed under the deceptively simple word *activation*. When we speak of a B cell being activated by an antigen, this encompasses all the changes in cell metabolism and gene expression which enable the cell to divide, differentiate into memory cells and plasma cells and for the latter to become factories making and secreting large quantities of antibodies.

4.7.1 Antibodies

The different types of antibody released by the plasma cells fall into five classes, termed G, A, M, D and E, distinguished by their immune functions. The antibodies in class G, for example, are termed *immunoglobulin G* (or IgG for short, pronounced 'eye-gee-gee'). Regardless of its class, each antibody has a basic structure consisting of four looping, interconnected polypeptide chains (see Figure 4.23; a simpler version is shown in Figure 4.22c). Two of these chains, the heavy (H) chains, are identical to each other and contain approximately 400 amino acids each. The other two chains, the light (L) chains, are also identical to each other, but only half as long as each H chain. The heavy chains have a flexible *hinge* region in their middle portion. The four chains combine to form an *antibody monomer* with two identical halves, each consisting of a heavy chain and a light chain. The molecule as a whole is T or Y shaped.

Each chain forming an antibody has a variable (V) region at one end and a much larger constant (C) region at the other end. Note that the regions are divided into smaller domains. Antibodies responding to different antigens have very different V regions, but their C regions are the same (or nearly so) in all antibodies of a given class. It is the V region which determines the antigenic specificity of the antibody. The V regions of the heavy and light chains in each arm of the monomer combine to form an antigen-binding site shaped to fit a specific antigen. The C regions that form the stem of the antibody determine the antibody class.

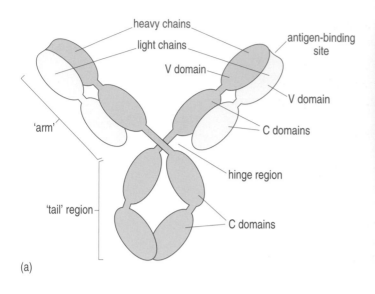

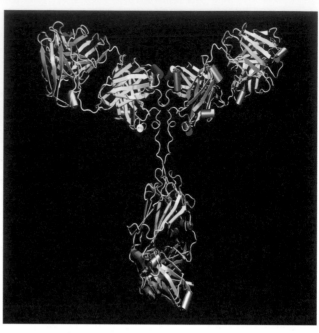

(a)

(b)

Figure 4.23 Antibody structure. (a) Basic antibody structure consists of four polypeptide chains linked together. Two of the chains are short light chains; the other two are long heavy chains. Each chain has a V (variable) region (different in different antibodies) and a C (constant) region (essentially identical in different antibodies of the same class). (b) Computer-generated image of antibody structure.

The ordering of the classes from G to E reflects their relative abundance in the bloodstream (G is the most abundant). The IgG class contributes about 85% of the antibody molecules in the circulation, whereas IgE is barely detectable. Whilst IgG may be the most abundant antibody class found in the bloodstream, it is not produced in significant quantities until the *secondary* adaptive immune response.

● Explain why more antibodies are produced during the secondary response than during the primary response.

○ The clonal expansion which takes place during the secondary immune response builds on an already enlarged clone of memory B cells, and so is far greater than the expansion possible during the primary response, which has only a few tens of cells as its starting point. In consequence, the amount of antibodies produced during the secondary response far exceeds that produced during the primary response.

The IgG antibodies in our circulation today therefore represent a sort of molecular record of the pathogens (and their epitopes) that we have encountered at least twice before in our recent past. These antibodies last longer than those of the other classes and can be detected for at least a month even after all traces of the triggering epitope have been eradicated from the body. IgG molecules have the useful property of being able to diffuse out of the bloodstream by passing through the walls of the thinnest blood capillaries. This extends their role into the tissues for a limited distance, particularly at sites of acute inflammation (described earlier in this chapter) where blood vessel walls become leaky and fluids and cells flood out. IgG antibodies are the only class which can pass through the blood vessels of the placenta and so transfer *passive immunity* from mother to unborn baby. IgG antibodies are also the most powerful of the *opsonins*, another term you met earlier.

● Can you recall what opsonins do, and explain how a molecule of IgG could fulfil this role? (If you need help, look back at Figure 4.7.)

● Opsonins form a bridge between targets (e.g. cellular pathogens or toxic macromolecules) and phagocytic leukocytes (e.g. macrophages), increasing the adherence between the two and promoting more efficient phagocytosis.

IgG antibodies also perform this bridging role for certain kinds of cytotoxic cells, which have greater efficiency in killing their targets if they first get a grip on antibodies bound to the targets. In effect, they act as potent target labels for leukocytes in the *innate* immune system, which themselves have relatively unsophisticated mechanisms for recognizing and binding to potential targets. The presence of IgG antibodies bound onto a cell or macromolecule (e.g. bacterial toxin) ensures its rapid destruction by the innate immune response. This is one of the ways in which the more sophisticated adaptive immune system builds on and focuses the older innate immune system.

IgG antibodies also have a role in activating one of the major chemical defence mechanisms – complement (Section 4.4.2). The first component in the cascade is strongly activated by contact with two or more adjacent IgG antibody molecules bound to a cellular target. This leads to the rapid scaling-up of the cascade reaction and the perforation of the target cell membrane by the final component of the complement (membrane attack complex). This is the *classical* pathway of complement activation as opposed to the *alternative* pathway mentioned in Section 4.4.2.

IgG antibodies have so many defensive functions that you may wonder why we need the other classes. The reason we do is because there are gaps in the immunological coverage provided by IgG. For a start, there is very little of it produced during a *primary* immune response to a novel antigen. The small number of B cells in the original clone differentiate into plasma cells which mainly synthesize antibodies of the class IgM. Molecules of IgM become joined together outside the cell in groups of five, forming a star-shaped structure called an IgM pentamer, a macromolecule so large that it can be photographed using an electron microscope (Figure 4.24a).

Figure 4.24 (a) Five molecules of IgM joined together in a star-shaped pentamer (magnified about 5 million times). (b) Schematic representation of the IgM pentamer shown in (a).

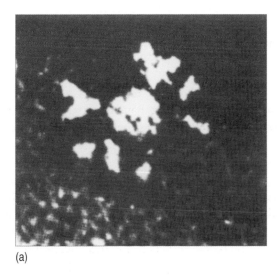

(a)

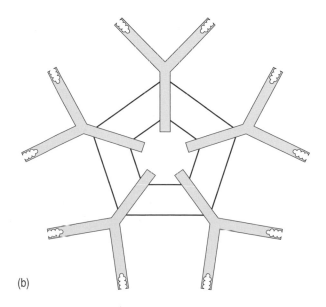

(b)

The ability of IgM molecules to form pentamers increases the effectiveness of the few available antibodies during the primary response to an antigen. Each pentamer has ten identical binding sites which makes it particularly good at *aggregating* antigens – causing the antigens to clump together. Clumps of antigens stuck together with IgM form sitting targets for the phagocytic and cytotoxic cells of the innate immune system. In addition, IgM antibodies are even more powerful activators of the complement cascade than are IgG antibodies.

However, both IgG and IgM have a major drawback: they can't get to parts of the body that are the most likely to be exposed to pathogens – the inner surfaces of the respiratory system (including the nose and mouth), the gut and the urinary and genital tracts. Most pathogens get into the body in the air we breathe, in the food and water we consume, and during sexual intercourse. This threat to health is countered by another class of antibody, IgA.

Although some IgA antibodies are found in the bloodstream, their unique feature is that they can be transported through the layer of cells lining our 'hollow' organs and become concentrated in the secretions there. IgA antibodies also occur in saliva and tears, and because of this location, are sometimes referred to as the first line of defence in the adaptive immune system. They can also link together in groups of two or three, but they cannot activate the complement cascade or cross the placenta. However, IgA does have a key role in the transfer of *passive immunity* from mother to new-born baby: it is the most abundant class of antibody in *colostrum*, the concentrated nutritional fluid produced by the breast in the first three days after birth and, to a lesser extent, in the subsequent breast milk.

IgD antibodies are found on the surface of B cells, however, their exact function is unclear and for the moment we shall say no more about them.

Finally, we come to the potential villain of the piece – IgE. The tail section of the IgE antibody has a specially adapted shape which enables it to bind to receptors on the surface of *mast cells*. If you think back to the earlier discussion of the acute inflammatory response, and look again at Figure 4.9, you should remember that when mast cells degranulate (release their packets of chemicals), they promote and enhance a rapid local inflammation. Several different triggers for mast cell degranulation are known, but one of the most potent involves IgE.

IgE antibodies usually have binding sites for epitopes found on common pathogens, including the larger multicellular parasites such as worms and flukes. During the primary response to these pathogens IgE antibodies are synthesized and become attached to mast cells. If a pathogen with epitopes that fit the IgE binding sites gets into the body for a second time, it will immediately trigger an acute inflammatory reaction around itself. IgE antibodies are one of the most important defences against parasitic infestation.

Unfortunately, excessive production of IgE antibodies can lead to health problems for a minority of people who synthesize IgE with binding sites that fit common but harmless proteins in our environment. Plant pollens, animal fur, house-dust mite faeces, bee and wasp venom and certain foodstuffs are the most prevalent of these allergens. Certain drugs (e.g. penicillin), plastics, rubber and metals can combine with proteins in the body to form potent allergens. In allergic individuals, once the IgE antibodies are formed and attached to their mast cells, any subsequent exposure to the allergen triggers mast cell degranulation and inflammation. Since the allergen usually enters the body via the airways or the digestive tract, or gets into the eyes, the allergic response most commonly affects these sites. In extreme cases, it can set off a chain reaction which spreads throughout the body, causing swelling and fluid build-up in the lungs and dilation of major blood vessels; the person may be unable to breathe and may suffer a fatal drop in blood pressure. This form of bodywide response is called **anaphylactic shock** (see also Section 2.6.5 and Figure 2.17). It typically occurs when the allergens directly enter the blood and circulate rapidly through the body, as might happen with certain bee stings or spider bites. It may also follow injection of a chemical such as penicillin. Many explanations abound as to why certain individuals are more susceptible to allergies than others. Some individuals inherit genetic traits that lead to overproduction of IgE by their plasma cells, but both environmental and developmental aspects are also involved.

We can now sum up the main ways in which secreted antibodies contribute to the immune response, in large measure by activating, focusing and enhancing the mechanisms of innate immunity (Figure 4.25).

● In all the examples summarized in Figure 4.25, the antibody molecule has to be *bound* to a complementary epitope on an antigen before it can activate or enhance an immune response. What is the value of this restriction?

● It ensures that *free* antibodies in the circulation do not set off an immune response indiscriminately; they can only do so in the immediate vicinity of the epitopes that fit their binding sites and hence they focus the response onto identified and (usually) legitimate targets.

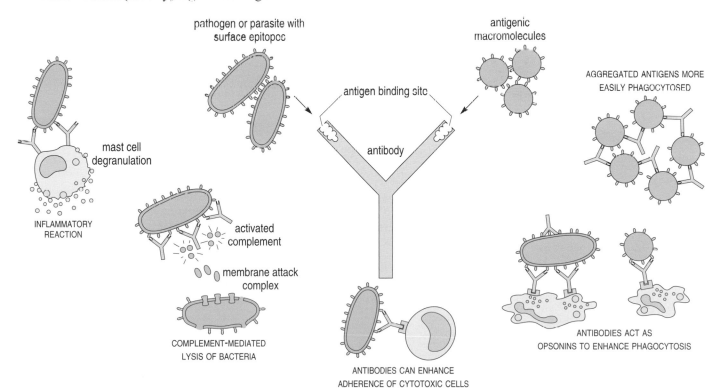

Figure 4.25 Summary of the main ways in which secreted antibodies contribute to the immune response against pathogens and their toxic products.

However, in case you get a little too enthusiastic about the defensive contribution of antibodies, it is worth noting that they are most effective against exposed targets. They do not easily reach the many types of pathogen that replicate and live inside the cells of their host – a group that includes the bacteria that cause tuberculosis, leprosy and Legionnaire's disease, the malarial parasite, and all the viruses, including the one that causes AIDS. Our main defence against intracellular pathogens are the NK cells and the cytotoxic T cells described earlier.

Summary of Section 4.7

1　The antigen receptors of B cells are called surface immunoglobulins, and are a surface-bound form of the antibody molecules which descendents of the activated B cell will eventually secrete.

2　After binding to their matching epitopes and exchanging activating signals (cytokines) with the corresponding clone of helper T cells, B cells differentiate into plasma cells; they secrete antibodies with binding sites for the same epitopes that initiated the immune response.

3　There are five classes of antibodies produced by plasma cells upon activation – IgG, IgA, IgM, IgD and IgE.

4　IgE antibodies cause degranulation of mast cells and are involved in acute inflammation. IgE antibodies are implicated in several forms of allergy.

5　B cells bind to intact *native* antigens, in contrast to the *processed* antigens bound by T cells.

4.8 Clonal deletion: the key to self-tolerance

Early in fetal life, when millions of different antigen-receptor shapes are being randomly generated as new lymphocyte clones are formed from the multipotent stem cells, a rigorous process of selection occurs in which self-reactive clones are weeded out and destroyed. This process is called **clonal deletion**. The majority of the clones that meet their deaths soon after they are formed in the fetus are *T cell clones*. Juvenile T cells are generated from stem cells by cell division in the bone marrow, but they then migrate to the thymus. Only a small minority of juvenile T cells entering the thymus emerge as mature T cells to join the pool of small lymphocytes circulating around the body – perhaps as many as 90% of them die in the thymus. This huge attrition is the consequence of programmed cell death, a process known as *apoptosis* (Book 1, Case Report 2.2).

The events inside the thymus which lead to the deletion of so many T cell clones are still somewhat mysterious, but the basic story can be summarized as follows. Specialized cells within the thymus present self epitopes on their surface membranes and, as the juvenile T cells pass through, any with antigen receptors that bind to the self epitopes are given a chemical signal which causes them to undergo apoptosis. The cell spontaneously shrinks, its nucleus collapses and the remains are removed by phagocytic cells. In the meantime, T cells with antigen receptors that do *not* bind to the self epitopes 'held out' to them in the thymus continue on their journey and emerge as mature T cells, each clone capable of binding to a different non-self epitope. By the time that development in the uterus is completed and the new individual is born, all of the T cells which could have bound to a self epitope have already been deleted and the immune system has become *self-tolerant*. The surviving T cells all have receptors that can bind only to non-self epitopes that get into the body for the first time *after* birth – usually as part of the structure of pathogenic organisms or their toxins.

The mechanisms of self-tolerance we have discussed so far have focused entirely on the deletion of self-reactive T cells. Clonal deletion is thought to have a relatively minor role in maintaining self-tolerance among the B cells; indeed, it is easy to demonstrate experimentally that plenty of self-reactive B cells are circulating around our bodies throughout our lives.

● Can you work out why these self-reactive B cells don't (in the majority of individuals) mount an immune response against the self epitopes which their antigen receptors can certainly bind to? (A glance back at Figure 4.21 may help.)

○ B cells require activating signals (cytokines) from helper T cells before they can mount an immune response, even when their own antigen receptor has bound to its matching epitope. The signals have to come from a helper T cell with antigen receptors that bind to an epitope on the same antigen as the B cell. Clonal deletion of self-reactive T cells ensures that self-reactive B cells are denied the activating signals they require to mount an immune response.

However, self-reactive B cells may become activated in later life, usually because new clones of self-reactive helper T cells are generated randomly and a few may slip through the net and survive the deletion process. Once activated, these 'anti-self' B cells manufacture antibodies that bind to specific epitopes in the body's own tissues. Self-reactive clones of cytotoxic T cells may also escape deletion. The damage that all these cells cause can be severe enough to produce an *autoimmune disease* (Book 1, Section 3.7.2; Book 2, Section 3.3.2). In all such diseases, the immune response is directed against a self epitope, but the symptoms vary depending on where in the host's body that epitope occurs. For example, autoimmune destruction of the cartilage in the joints occurs in **rheumatoid arthritis** (see also Book 2, Case Report 4.2), and in autoimmune thyroiditis the person's thyroid gland is attacked. Autoimmune responses can be directed against a wide variety of self epitopes, including some found in the structure of DNA!

Autoimmune disease can afflict people of all ages. Case Report 4.2 tells the story of Alice, a young and promising tennis player (Stuttaford, 2004).

Case Report 4.2   Rheumatoid arthritis

In 1992 Alice Peterson was playing tennis in national tournaments and aiming for Wimbledon. She was 18 and about to take up a tennis scholarship in America. One morning while competing in a tournament she discovered that she couldn't grip her racket. Her thumb joint was slightly swollen but not painful, just weak.

She assumed that it had been the result of a trivial injury she had sustained the day before and went out on to the court. She picked up her racket but couldn't hold it. Tennis was out. She was sent to the local A&E, had the thumb strapped, was given some mild painkillers and told to go home. The next day she was playing tennis with all her killer ambition and soon dismissed all thoughts of her hand until 24 hours later the swelling and weakness returned.

Her life had changed for ever. Within a week or two the fleeting aches and pains that darted from joint to joint had become constant, Alice was beginning to feel lousy, her movements had slowed and her energy had disappeared. Alice had rheumatoid arthritis.

American scholarships were out, her Wimbledon ambitions had evaporated and she couldn't even keep going at Bristol University, her alternative to the tennis circuit.

A hip replacement eventually enabled her to return to Bristol and gain a first-class degree but her physical life was still crippled and her spirit was becoming crushed by the pain. She had tried all the major treatments for rheumatoid arthritis but the only

answer seemed to be surgery to replace destroyed joints as one by one they crumbled.

A few years ago she was included in a trial of [a new drug, one of the group of drugs that inhibit the action of the cytokines released by helper T cells and macrophages]. She is revitalised, pain is under control, mobility is restored (after yet more surgery) and she is revelling in family life and friendships. She has written two books and says: 'For me the effect of the [anti-cytokine] drugs has been magical.'…

4.9 Immunity: active and passive

To end this chapter we will discuss more fully what is meant by active and passive immunity. We have met several examples of both passive and active immunity as this chapter unfolded. **Active immunity** refers to resistance to a specific pathogen acquired as a result *either* of an earlier infection with that pathogen, from which the host has subsequently recovered, *or* through **immunization** – the deliberate introduction into the body of dead or live artificially weakened microbes (vaccines), or deactivated toxins (toxoids).

● From what you now know about adaptive immunity, what do you think a vaccine contains, and what is the rationale for immunization?

● Vaccines contain the unique epitopes found in the structure of, or secreted by, a particular pathogen. (Killed whole organisms may be used, or organisms that have been rendered harmless, or which have very similar epitopes to the pathogenic species; some vaccines contain inactivated toxic molecules that the pathogen secretes, and increasingly, synthetic epitopes constructed by protein biochemists are being tested.)

The rationale of immunization is that exposing an individual to the 'safe' epitopes in the vaccine for the first time will elicit a primary adaptive response without the person becoming ill. If the same epitopes are encountered subsequently on the live pathogen, the enhanced secondary response will be effective enough to eliminate it causing, at worst, only mild symptoms of the disease.

The vaccines and toxoids retain the antigenic properties that stimulate the development of immunity but they cannot cause the disease. 'Active' signifies that the person concerned has mounted an immune response against the pathogen, vaccine or toxoid, which has led to a state of ongoing immunity. Many microbial diseases can be prevented by artificial immunization. Examples are shown in Table 4.2.

Active immunization against some infectious diseases will confer life-long immunity, e.g. diphtheria, whooping cough or mumps. In other infections the immunity may only last for a number of years or for only a few weeks before

Table 4.2 Diseases preventable by vaccination.

anthrax	rubella
cholera	small pox
diphtheria	tetanus
hepatitis B	tuberculosis
measles	typhoid
mumps	whooping cough
poliomyelitis	

revaccination is necessary. Apparent loss of immunity may be due to infection with a different strain of the same microbe, which has different antigenic properties but causes the same clinical illness. The viruses that cause the common cold and influenza fall into this category. In the elderly and when nutrition is poor the production of lymphocytes, especially B lymphocytes, is reduced and the primary and secondary response may no longer be adequate.

By contrast, **passive immunity** refers to a temporary resistance to a specific pathogen acquired as a result of receiving the products of *someone else's* active immune response. As mentioned above, throughout this chapter we have occasionally referred to forms of immunity derived passively.

● Can you recall two examples of situations in which passive immunity is acquired early in life?

● Examples are:
   • The fetus passively acquires some of the antibodies actively produced by the mother in response to infections she has encountered; in the placenta these antibodies pass from the mother's bloodstream into the fetal circulation.
   • The new-born baby who is breast-fed also passively acquires some maternal antibodies in the colostrum that the mother secretes in the first three days after giving birth and, to a lesser extent, in the breast milk thereafter.

Summary of Sections 4.8 and 4.9

1   Self-tolerance is achieved during fetal development by the clonal deletion of T cells with antigen receptors that fit self epitope.

2   Antibodies play a key role in adaptive immunity; they bind to antigens, labelling them for destruction.

3   The adaptive immune system responds specifically to each unique antigen and is normally self-tolerant.

4   Antibodies and other products of an active immune response can confer passive immunity on another individual.

4.10 Bodily interactions

The immune system does not exist in isolation but is an integral part of the complex machinery that we call our body. One of the most interesting recent developments in immunology has been the recognition that some of the cytokines synthesized and secreted by leukocytes during an immune response have exactly the same molecular structures as some of the signalling molecules found in the nervous system (these include certain neurotransmitters, discussed in Section 1.9.2 of Book 2). Without realizing it, immunologists, endocrinologists (Book 2, Section 3.8) and neuroscientists have been studying the same molecules in different sites in the body and have given them different names. Clearly, these 'shared role' molecules produce different effects if the cell receiving the signal is a neuron rather than (say) a macrophage, but their existence has lent weight to the growing body of evidence that the nervous system and the immune system are in constant communication.

Cells in the immune system have molecules on their surfaces which are receptors for (and hence bind to) neurotransmitters and other signalling molecules released by the nervous system. Every lymphatic organ (e.g. the lymph nodes, spleen, tonsils, etc.) has an extensive network of nerve fibres penetrating the entire structure. These findings suggest that the nervous system can control the activity of the immune system and, although the precise mechanisms are not yet clear, the extent of the influence is thought to be considerable. Psychological interactions are also important: for example, laboratory rats and mice can be trained by a system of rewards and unpleasant stimuli to enhance or inhibit their immune responses to antigens.

Experiments such as these raise the possibility that human immune response may be affected by such imponderable phenomena as 'mood' and 'stress' – a subject to which we return in Book 4, Chapter 3 of this course. Following periods of stress, most animal studies have shown alterations in the levels of antibodies in the circulation and in the total number of leukocytes in the blood. These changes are in the direction of *reducing* the immune system's capacity to respond quickly and effectively to infection. Evidence that *people* in relatively common stressful situations (e.g. taking exams or moving house) do indeed succumb to a greater number, or more prolonged episodes, of infection has not always been convincing, but there is a general consensus that they do – although the deficit in the immune responsiveness is thought not to be great enough to be life-threatening as a general rule.

The recognition that the immune system is part of a greater interacting network of communication and response in the body has been most clearly established in relation to the endocrine system. Corticosteroid hormones secreted by the adrenal glands close to the kidneys, under the control of the hypothalamus and the pituitary gland in the brain, can have a significant *suppressive* effect on the immune system. Under normal conditions, these hormones are believed to be part of the fail-safe mechanisms that switch off an immune response. If the output of these hormones exceeds a certain threshold, however, as it does when a person is faced with a very stressful situation, a more generalized immunosuppression occurs. At first sight, this outcome seems counterproductive. However, the immune systems uses a huge amount of energy when it is activated by a new infection, so there appears to a trade-off between the short-term gain of saving energy for fight or flight in the face of potential danger, at the expense of a short-term reduction in immune responsiveness to an infection that might occur (but see also Chapter 3 of Book 4).

It is clear that the nervous, endocrine and immune systems are continuously interacting and modulating each other's activity. Some of the most compelling demonstrations of this interactivity have come from studies of the effects of exercise on the immune system. Elite athletes in rigorous training for a major event have an increased susceptibility to infection, particularly of the respiratory tract.

Whilst excessive or unaccustomed exercise has marked suppressive effects on several components of the immune system, the good news for most of us is that the immune system usually recovers its normal values within six hours. Indeed, regular moderate exercise gradually raises the normal levels of circulating leukocytes, which may result in a moderate increase in immune responsiveness.

Questions for Chapter 4

Question 4.1 (LOs 4.1, 4.2, 4.4)

Give three examples of ways in which small lymphocytes engaged in an adaptive immune response increase the speed and efficiency of antigen destruction by affecting the activity of leukocytes in the innate immune system.

Question 4.2 (LO 4.1)

In what ways can the adaptive immune response be said to be partly dependent for its activation on cells belonging to the innate immune system?

Question 4.3 (LO 4.4, 4.6, 4.7)

Would you expect a person who has recovered from measles to show less susceptibility to chickenpox infection than a person who has never had measles? Explain your answer.

Question 4.4 (LO 4.4, 4.6, 4.7)

Explain why infectious diseases common among children in some countries (such as measles, which is caused by a viral infection) do not generally occur twice in the same person. Explain your answer in terms of clonal selection theory.

Question 4.5 (LO 4.6)

The colostrum produced by the mammary glands in the first few days after birth contains antibodies with binding sites for any antigens that the mother has encountered in recent weeks. Antibodies are also found to a lesser extent in her breast milk. What features of the immune system of a new-born baby make these passively acquired antibodies so important for its health? What immunoglobulin class do the antibodies in colostrums and breast milk primarily belong to, and why is this significant?

Question 4.6 (LO 4.5)

Distinguish between self-MHC restriction and self-tolerance as properties of the adaptive immune system.

Question 4.7 (LO 4.5)

How is self-tolerance primarily established and what mechanisms cause it to break down?

Question 4.8. (LO 4.3)

Explain how an acute inflammatory response can contribute both to protection from pathogens and to allergic reactions to harmless environmental proteins.

References

Boccaccio, G. (1921, first written in 1349–51) *The Decameron* (trans. J. M. Rigg), London: David Campbell.

Gallon, D., 2003 [online] Available from: http://www.nnff.org/survivors/dan_gallon/dan_gallon.htm. (Accessed January 2005)

Stuttaford, T. (2004) 'A helping hand for arthritis', *The Times* (*Times 2*), 5 February, p. 12.

CONCLUSION

We earlier made mention of the practical necessity of studying biology, system by system, in this course and explained that this would result in our adopting a reductionist approach most of the time. As you accumulated more knowledge of the different body systems it became possible to make the links between the systems more explicit; indeed it would have been impossible to discuss the homeostatic mechanisms involved in the regulation of blood pressure, for example, in terms of the activity of the cardiovascular system alone.

The last section of Chapter 4 was a reminder of the holistic theme that we have tried to maintain sight of throughout the course. Our mental life, moods and emotions, are grounded in our biology. We use our brains when we think, requiring the synthesis and release of neurotransmitter molecules; molecules that, as you have just read, can also affect the immune system. The idea that 'laughter is the best medicine' is hardly new but the extent of laughter's physiological effects on blood pressure, heart rate, breathing and muscular tension as well as its effects on the immune system and levels of certain neurochemicals are still being researched.

Stress might be regarded as the other side of the coin to laughter. Yet physiological stress responses are essential for our survival in dangerous or challenging situations, and also enable us to perform demanding tasks, such as speaking before an audience or taking part in competitive sport. Once the stressful situation has come to an end, homeostatic mechanisms come into play to smooth out the physiological effects of the stress response and restore homeostasis. Sleep plays an important part in restoring us to what feels like a comfortable state, but is in fact a state of homeostasis. Yet as we shall see in Book 4 prolonged stress and consequent maintenance of a physiological response to stress have harmful long-term effects.

If we were to say that you have now studied the body systems and the way they interact to maintain homeostasis and enable individuals to live happy, productive lives, you might protest that we have yet to make mention of the reproductive system. Clearly sexuality is part of our biology, but is it necessary; does it have any homeostatic function? You will probably be aware that there are organisms which are not sexually differentiated (i.e. they do not require two distinct forms – male/female – for reproduction to occur). Instead they can reproduce by splitting into two; a form of asexual reproduction. So it seems that sexuality is not necessary and reproduction without sexuality is perfectly possible. However, in the first chapter of Book 4 we will be arguing that there are particular benefits to be gained from a sexual mode of reproduction.

● What do you think is the answer to the question raised above? Does reproduction have a homeostatic function?

● Many individuals live their lives without either reproducing or performing any sexual activity that could result in reproduction. Thus we must conclude that reproduction does not serve any homeostatic purpose for the individual.

Once born the only certainty is that we will die. This is a bit of a problem for *individuals* as we are incapable of reproducing ourselves by cloning. (As you read through Book 4 you might like to think why this should be so.) Nor is there any certainty that we will reproduce, and as you know from Section 1.4 of Book 1, an individual that dies without reproducing has no biological fitness. If all the individuals of our species were to die without issue that would, patently, be the end of the species. *Homo sapiens* would become extinct. Now, homeostasis is a term that is applied to describe individual physiology; it is not used in connection with the 'health of the species'. However, the role of reproduction is crucial to the maintenance of population size from one generation to the next and thus it has a role that is analogous to a homeostatic one. Without question we cannot draw to a close our study of human biology without studying the reproductive system.

So in the final book of this course we consider the tremendous diversity between individuals and explain how this comes about and we look at some of the changes and challenges that we face in our adult lives.

ANSWERS TO QUESTIONS

Question 1.1

Substance X may be filtered at the glomerulus, and then partially reabsorbed. Alternatively it may not be filtered at all, but secreted into the filtrate (Equation 1.1):

$$\text{clearance of X/ml per min} = \frac{\text{amount of X excreted/mg per min}}{\text{plasma concentration of X/mg per ml}}$$

$$\text{clearance of X/ml per min} = \frac{2 \text{ mg per min}}{8 \text{ mg per ml}} = 0.25 \text{ ml per min}$$

Question 1.2

(a) Glucose is reabsorbed by active transport coupled to sodium, in the proximal tubule. (b) Sodium is reabsorbed by active transport, coupled to a number of other substances, including glucose, amino acids, bicarbonate and phosphate. It is also absorbed by diffusion through protein channels. (c) Water is reabsorbed passively, by osmosis through protein channels in the membranes of the epithelial cells lining the kidney tubules, and by diffusion through the small spaces between these cells. In the collecting ducts, the number of protein channels through which water is absorbed is increased by ADH.

Question 1.3

See Figure 1.29. The regulated variables are the body fluid levels of water and sodium.

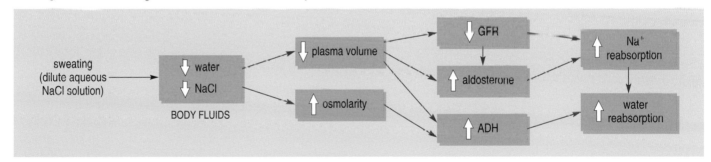

Figure 1.29 Answer to Question 1.3.

Question 1.4

It will result in acidosis, as, in Equation 1.2, there will be a shift to the right, and hence an increase in the concentration of hydrogen ions. Under these circumstances, the amount of hydrogen ions secreted will be greater than the quantity of bicarbonate filtered, and hydrogen ions will therefore be excreted to restore pH balance.

Question 1.5

A reduction in oxygen concentration stimulates the release of the hormone erythropoietin from capillary endothelial cells in the kidney. Erythropoietin, which is also produced by the liver, stimulates the proliferation of red blood cell precursors in the bone marrow.

Question 2.1

(a) Haematocrit ratio is the volume fraction of a sample of blood occupied by packed erythrocytes. The haematocrit ratio is expressed as a percentage.

(b) Vagal restraint refers to the effect that the continuous inhibitory cholinergic input derived from the vagus nerve (10th cranial nerve) has on the activity of the sinoatrial node (SAN) in the heart.

(c) Myocardial ischaemia is defined as a reduction or complete lack of blood flow through blood vessels supplying part of the heart.

(d) The blood–brain barrier restricts the movement of ionic solutes into the substance of the brain. The internal environment of the brain is therefore tightly controlled and the neurons (and other brain cells) are protected from fluctuating levels of ions, hormones and tissue metabolites circulating in the blood plasma.

(e) A drug antagonist is a compound that specifically competes with a drug for access to a receptor molecule or recognition site. The antagonist prevents or interferes with drug–receptor interactions and therefore the action of the drug on target tissues.

(f) An inotropic cardiac drug indirectly increases the intracellular concentration of Ca^{2+} ions in heart cells which increases the force of cardiac muscle contraction.

(g) A thromboembolism is a freely moving blood clot that has the potential to either partially or completely block blood vessels anywhere in the body, including the heart, lungs and brain.

(h) Orthostatic hypotension is a temporary condition where low blood pressure occurs due to inadequate circulatory compensation to the gravitational shifts in blood distribution when an individual suddenly moves from lying down to a sitting or standing position.

Question 2.2

(a) As an erythrocyte in the blood, you enter the right atrium via the inferior or superior venae cavae and pass through the tricuspid (atrioventricular valve) into the right ventricle. (Atrial contraction ensures that you are pumped out of the atria and into the ventricles.) You are then pumped from the right ventricle into one of the two pulmonary arteries and travel to the lungs. Your backflow into the right side of the heart is prevented by closure of the pulmonary semilunar valve. You return from the lungs to the heart via one of the four pulmonary veins into the left atrium. From the left atrium you pass through the bicuspid (mitral) valve into the left ventricle and are subsequently pumped out of the heart into the aorta and then around the body. You are prevented from flowing back into the left side of the heart by closure of the aortic semilunar valve.

(b) The sinoatrial node (SAN, a group of specialized cardiac muscle cells) is the heart's pacemaker and generates rhythmical waves of electrical excitation which spread through the atrial muscle cells, each wave leading to contraction of the atria. The electrical activity reaches the atrioventricular node (AVN), where the speed of conduction slows temporarily, allowing the atria to complete their contraction before the ventricles themselves contract. The conduction speed increases again when the electrical impulses reach the bundle of His and is conducted along the Purkinje fibres to the base of the ventricles. Ventricular contraction follows the electrical depolarization of the ventricles, and blood is expelled as the wave of contraction moves from the base of the ventricles upwards through the ventricular walls.

Question 2.3

The normal adult ECG is generally composed of a P wave, a QRS complex and a T wave. The normal ECG has a characteristic and regular waveform produced by synchronized patterns of electrical activity responsible for the activity of the various parts of the heart. In a normal ECG, a P wave (representing atrial depolarization) precedes each QRS complex (ventricular depolarization and atrial repolarization); these are then followed by a T wave (ventricular repolarization). In a person with 3 : 1 AV heart block, only every third atrial impulse is transferred to the ventricles to produce one ventricular beat. Consequently, the ECG of a person with 3 : 1 AV heart block shows on average three times as many P waves occurring for each QRS complex. The P waves occur irregularly and at any stage of the cardiac cycle.

Question 2.4

(a) Using the relationship:

$$\text{cardiac output} = \text{heart rate} \times \text{stroke volume}$$

$$\text{cardiac output} = 70 \text{ beats per min} \times 70 \text{ ml}$$

$$= 4900 \text{ ml per min}$$

$$= 4.9 \text{ litres per min}$$

$$= 4.9 \times 60 \text{ min} = 294 \text{ litres per hour}$$

(b) Cardiac output can be altered by changes in heart rate or stroke volume or both, as given by the equation used in (a). Heart rate is the major controlling factor and it is increased by increased sympathetic (noradrenergic) nervous activity stimulating the sinoatrial node (SAN); it is decreased by parasympathetic (cholinergic) innervation of the SAN derived from the vagus nerve. The heart rate can also be influenced by humoral factors (e.g. circulating adrenalin and noradrenalin – the 'fright, fight and flight' hormones). Stroke volume is dependent on the venous return (Starling's law), so an increase in venous return (pre-load) will increase stroke volume. Contraction of the smooth muscle in the walls of the veins and venules increases the venous pressure and forces a greater volume of blood back towards the heart, so increasing venous return.

Question 2.5

Blood pressure measurements are given as 'systolic/diastolic'. Systolic pressure refers to the maximal arterial blood pressure when the ventricles have fully contracted and have expelled the blood contained inside into the systemic circulation. In contrast, diastolic pressure refers to the minimal arterial blood pressure when the ventricles have fully relaxed and are filling with blood.

A patient with a systolic pressure exceeding 160 mmHg and a diastolic pressure above 95 mmHg is considered to have high blood pressure or hypertension.

Hypertension can be treated by drugs that reduce cardiac output and/or the total resistance of the peripheral circulation (blood pressure = cardiac output × peripheral resistance). The major classes of antihypertensive drugs include: β-adrenoceptor blockers (e.g. metoprolol), which act to decrease cardiac output; diuretics (e.g. thiazide and loop diuretics), which reduce blood volume; angiotensin-converting enzyme (ACE) inhibitors (e.g. captopril), which prevent the synthesis of the potent vasoconstrictor angiotensin II; and vasodilating drugs (e.g. prazosin), which reduce peripheral resistance by selectively blocking the constrictive effects of α1-adrenoceptor stimulation.

Vasodilation can also be produced by drugs (e.g. nifedipine) that cause a relaxation of arteriolar smooth muscle by blocking entry of Ca^{2+} ions into cells through calcium channels in the muscle cell membranes.

Question 3.1

(a) The oxygen–haemoglobin dissociation curve is an important physiological index demonstrating graphically the interaction between haemoglobin and O_2 in the blood. The graph plots the amount of oxyhaemoglobin present (as a percentage of the total haemoglobin) versus the partial pressure of O_2 in the blood.

(b) Pleurisy is an inflammation of the pleural membranes, which is caused by infection of the pleural cavity.

(c) Peak flow rate is the maximal rate at which air can be expelled through the conducting vessels of the lungs. It is usually measured during the first 100 milliseconds of forced expiration from normal total lung volume and the flow rate is expressed in litres per minute.

(d) Carboxyhaemoglobin (HbCO) is the name of the compound formed by the irreversible combination of carbon monoxide and haemoglobin: $Hb + CO \rightarrow HbCO$.

(e) The Bohr shift occurs when blood becomes slightly more acidic than normal (i.e. less than pH 7.4), as occurs in respiring tissues. This causes the normal oxygen–haemoglobin dissociation curve to 'shift' laterally to the right, which results in a decreased affinity of haemoglobin for O_2, so that O_2 is more readily given up to the tissues.

(f) Sinus arrhythmia is the normal variation in heart rate during breathing. Heart rate goes up slightly as we breathe in and down when we breathe out.

(g) The diving response is an innate physiological reaction to complete immersion of the face and body in water: the heart rate slows and blood is only pumped to vital organs, so thereby conserving the limited supply of oxygen.

Question 3.2

Air enters the body through the nose or mouth and passes via the larynx into the trachea. The trachea branches into two bronchi which divide repeatedly into progressively narrower airways: smaller bronchi, bronchioles and terminal bronchioles, alveolar ducts and finally alveoli. Contraction of the intercostal muscles causes an upward and outward movement of the ribs, and contraction of the diaphragm further increases the volume within the chest cavity. Since the pleural membranes covering the lungs are held against the chest wall by surface tension, the lungs are forced to expand by the expansion of the chest wall. Air rushes into the expanded lungs to fill the increased volume inside the chest cavity. This activity is called inspiration.

Question 3.3

Oxygen is carried in the blood in combination with the respiratory pigment haemoglobin inside the erythrocytes. Oxygen and haemoglobin combine (reversibly) to form oxyhaemoglobin in erythrocytes (see equation below). Oxygen is transported in this form in the blood from the lungs to the sites of utilization (e.g. exercising muscles).

$$Hb + 4O_2 \rightleftharpoons Hb(O_2)_4$$

haemoglobin   oxyhaemoglobin

Carbon dioxide is carried in the blood either physically dissolved (10%), bound to haemoglobin (30%) or as bicarbonate – in fact, the majority (60%) of CO_2 is carried in the blood as bicarbonate ions (HCO_3^-; see equation below). Carbon dioxide is not very soluble in blood and can only bind weakly to the globin portion of haemoglobin. The first step in the conversion of CO_2 to HCO_3^- ions is catalysed by the enzyme carbonic anhydrase which accelerates the reaction between CO_2 and H_2O (see equation below); the product is carbonic acid (H_2CO_3) which dissociates into HCO_3^- ions and H^+ ions:

$$CO_2 + H_2O \underset{\substack{\text{carbonic} \\ \text{anhydrase}}}{\rightleftharpoons} H_2CO_3 \rightleftharpoons HCO_3^- + H^+$$

carbon dioxide   water   carbonic acid   bicarbonate ion   hydrogen ion

The H^+ ions are buffered biochemically (neutralized) by haemoglobin, and the HCO_3^- ions diffuse out of the cell into the plasma. An inward movement of chloride ions (Cl^-), known as the 'chloride shift', compensates for the loss of HCO_3^- ions from the erythrocytes.

Question 3.4

The commonly prescribed drugs to treat asthma are bronchodilators, e.g. salbutamol or salmeterol. These drugs act by stimulating sympathetic β2-adrenoceptors in bronchial smooth muscle to cause bronchial dilation which produces a decrease in airway resistance – such drugs are called β2-adrenoceptor agonists. In severe acute asthma attacks, additional drugs such as iprotropium bromide (that block the bronchoconstrictive action of parasympathetic acetylcholine) are also given. The effect of these bronchodilator drugs result in an increased air-flow through the bronchioles, which greatly eases the effort of expiration and relieves the symptoms of asthma. Corticosteroid drugs, such as inhaled beclomethasone, are used as regular long-term preventative (prophylactic) agents. These drugs reduce bronchial mucosal inflammatory reactions (e.g. oedema production and excess mucus secretion) by modifying cellular reactions to potential allergens.

Question 3.5

During exercise, the working muscles require an increased supply of O_2, which demands simultaneous increases in cardiac output and ventilation. An increase in heart rate and stroke volume increases cardiac output. The heart rate is increased by a reduction in the tonic inhibition of the sinoatrial node. Stimulation by the sympathetic nervous system and the action of circulating adrenalin and noradrenalin cause vasoconstriction, which increases the peripheral resistance and venous return (pre-load), which by Starling's law results in increased stroke volume and thus cardiac output.

Sympathetic nerve activity also increases ventilation by stimulating the contraction of the inspiratory muscles. Metabolic autoregulation in the muscle capillary beds increases the local blood supply through the tissues. A build-up of lactate during exercise shifts the oxygen–haemoglobin dissociation curve to the right (see Figure 3.6) with the result that O_2 is released from haemoglobin more easily. After the cessation of exercise, the alterations in ventilation and cardiac output persist until the post-exercise oxygen consumption caused by lactate build-up is returned to a normal level.

Question 4.1

There are many examples, but here are three (you may have thought of others).

1   Of greatest importance are the cytokines produced by helper T cells, which attract, activate and retain phagocytic and cytotoxic leukocytes at the scene of an immune response.

2   Activated B lymphocytes differentiate into plasma cells which secrete antibodies. These can act as opsonins, i.e. they form a bridge between phagocytic cells and their targets, thereby increasing the clearance of antigens by phagocytosis.

3   Antibodies bound to target cells also label them for destruction by the cytotoxic cells of the innate immune system.

Question 4.2

T cells can only bind to epitopes which are presented to them in the cleft of a self-MHC molecule on the surface of another cell. Correct antigen presentation to helper T cells is essential for T cell activation, and hence for the production of all the helper cytokines which activate or enhance the activity of all the other cells in the immune response. Although B cells can present antigen to T cells, so too can macrophages and various other cells that belong to the innate immune system. In addition, macrophages produce a cytokine of their own, which is essential for T cell activation.

Question 4.3

A person who has recovered from measles is just as likely to catch (say) chickenpox for the first time as a person who has *never* had measles. The immune system shows *antigen specificity*, i.e. the ability to *recognize* each *antigen* specifically in terms of its unique *epitopes*, so the immune system of a person who has recovered from measles is *not* adapted to respond more effectively against any other pathogen. The clonal expansion that accompanied the measles infection only involves those clones of small lymphocytes with *antigen receptors* that bind measles epitopes.

Question 4.4

When a child becomes infected with (say) the virus that causes measles for the first time, members of any clones of small lymphocytes in the adaptive immune system which have *antigen receptors* or the correct shape bind to *epitopes* on the virus (*antigen recognition* leads to *clonal selection*). Clonal expansion follows and various types of defensive cells differentiate and initiate a primary immune response directed specifically against the measles virus epitopes. After the virus has been eliminated, the immune system remains permanently adapted to mount a faster, more effective *secondary response* against the measles virus if the same epitopes are ever encountered again. This ability (*immunological memory*) relies on the differentiation of long-lived memory cells during the *primary response*, which form an expanded clone of cells ready to react quickly and effectively to a second exposure to the measles epitopes. It is unusual to have measles twice because the person usually becomes *immune* to the virus after the first exposure.

Question 4.5

New-born babies have no memory cells in their circulation because (ideally) they will not have encountered any antigens while in the protection of the womb. So all of their clones of small lymphocytes are 'unexpanded' and each consists of only a few tens of cells. As a result, the adaptive immune response of a baby to a new antigen will be the slow and less-effective primary response, which may be unable to eradicate an infection before symptoms develop. Passively acquired antibodies from colostrum and breast milk are therefore an extremely important defence against infection, particularly against pathogens entering through the baby's gut. This is partly because antibodies arrive first in the gut via colostrum and breast milk, but they are primarily of the IgA class, which concentrates in the lining of the gut and gives immediate protection there.

Question 4.6

Self-MHC restriction refers to the inability of T cells to bind to their complementary epitopes unless they are presented in the cleft of a 'self' MHC molecule on the surface of one of the body's own cells. *Self-tolerance* refers to the selective unresponsiveness of the adaptive immune system to self epitopes, i.e. those molecular configurations which occur in the structures of the cells and macromolecules of the individual's own body.

Question 4.7

Self-tolerance is primarily established through the deletion of self-reactive clones of T cells as they mature in the thymus. In the absence of helper T cell cytokines, self-reactive clones of B cells can survive but cannot be activated by contact with antigen alone. If a self-reactive clone of helper T cells escapes deletion, then it can activate the pre-existing clone of self-reactive B cells which then differentiate into plasma cells producing anti-self (autoimmune) antibodies. Self-reactive clones of cytotoxic T cells may also be generated and if these escape deletion they will mount attacks against the body's own normal cells.

Question 4.8

The acute inflammatory response can be triggered by several different stimuli, including the unique sugar and proteins in the cell walls of certain pathogens, and by the activated components of the complement cascade. One of the most important triggers occurs when antibodies of the IgE class bind both to the epitopes on a pathogen and to the surface of mast cells. This event causes the mast cells to degranulate, emptying their chemically reactive granules onto the pathogen. However, some individuals synthesize IgE antibodies with binding sites that fit common but harmless environmental proteins (allergens); if these IgE antibodies become bound to mast cells, then further contact with the allergen triggers an acute inflammatory response. The tendency to develop allergies seems to rely partly on the inherited over-production of IgE. It may also rely on the failure of the immune system to mature correctly if the early environment supplies high exposure to common allergens coupled with low exposure to common pathogens.

ACKNOWLEDGEMENTS

Grateful acknowledgement is made to the following sources for permission to reproduce material within this product.

Figures

Figure 1.2: PA Photos; *Figure 1.6b*: Professor P. M. Motta and M. Castellucci/ Science Photo Library; *Figure 1.7* (bottom right), *Figure 1.25b*: CNRI/Science Photo Library; *Figure 1.9*: Adapted from Marieb, E. N. (2004) *Human Anatomy and Physiology*, International Edition, copyright © 2004 Pearson Education, Inc.; *Figure 1.10*: Damien Lovegrove/Science Photo Library; *Figure 1.14a*: SIU/ Science Photo Library; *Figure 1.14b*: Dr M. A. Ansary/Science Photo Library; *Figure 1.16*: Krane, C. M. and Kishore, B. K. (2003) 'Aquaporins: the membrane water channels of the biological world', *Biologist*, **50** (2); *Figures 1.21, 1.22, 1.23*: Adapted from Vander, A. J., Sherman, J. H. and Luciano, D. S. (2001) *Human Physiology, The Mechanisms of Body Function*, 8th international edition, copyright © 2001, 1998, 1994, 1990, 1985, 1980, 1975, 1970 by McGraw-Hill, Inc., with permission of The McGraw-Hill Companies; *Figure 1.25c*: Science Photo Library; *Figure 1.27*: BSIP Laurent/H. Americain/Science Photo Library.

Figures 2.2, 2.5, 2.8: Adapted from Claude A. Vilee (1989) *Biology*, 2nd edn, copyright © 1989 by Saunders College Publishing; *Figure 2.4b*: Science Photo Library; *Figure 2.7*: Ganong, M. D. F., Lange, J. and Lange, D (1983) *Review of Medical Physiology*, 11th edn, Appleton and Lange, by permission of Professor W. F. Ganong MD and the publisher.

Figures 3.2 and 3.3: Adapted from Claude A. Vilee (1989) *Biology*, 2nd edn, copyright © 1989 by Saunders College Publishing; *Figure 3.5*: van Wynsberghe, D., Noback, C. R. and Carola, R. (1995) *Human Anatomy and Physiology*, 3rd edn, McGraw-Hill, Inc., by permission of McGraw-Hill Companies; *Figure 3.6*: Ganong, W. F., Lange, J. and Lange, D. (1983) *Review of Medical Physiology*, 11th edn, Appleton and Lange, by permission of Professor W. F. Ganong MD and the publisher; *Figure 3.9*: The Wellcome Trust.

Figures 4.1a, 4.8, 4.9a, b, 4.13, 4.23b: Science Photo Library; *Figure 4.1b*: University of South Carolina; *Figure 4.5*: Courtesy of Professor Robert Dourmashkin; *Figure 4.6*: Courtesy of Dr P. M. Lydyard; *Figure 4.13*: Science Photo Library; *Figure 4.15 a, b*: Courtesy of Dr W. van Ewijk; *Figure 4.24*: Feinstein, A. (1971) *Annals of the New York Academy of Sciences*, **190**, Plate 4.2.

Tables

Table 1.3: Vander, A. J., Sherman, J. H. and Luciano, D. S. (2001) *Human Physiology, The Mechanisms of Body Function*, 8th international edition, copyright © 2001, 1998, 1994, 1990, 1985, 1980, 1975, 1970 by McGraw-Hill, Inc., with permission of The McGraw-Hill Companies.

Tables 4.1 and 4.2: Reprinted from Waugh, A. and Grant, A. (2001) *Anatomy and Physiology in Health and Illness*, 9th edn, copyright © 2001, with permission from Elsevier.

Text

Case Report 4.1: Dan Gallon's Survivor Story, by permission of Dan Gallon;
Case Report 4.2: 'A helping hand for arthritis', *The Times*, 5th February 2004.

Every effort has been made to contact copyright holders. If any have been inadvertently overlooked the publishers will be pleased to make the necessary arrangements at the first opportunity.

INDEX

Entries and page numbers in **bold type** refer to key words which are printed in **bold** in the text. Page numbers in italics are for items mainly or wholly in a figure or table.